Best regards,

[illegible]

SILENCING THE BOMB

LYNN R. SYKES

SILENCING THE BOMB

One Scientist's Quest
to Halt Nuclear Testing

COLUMBIA UNIVERSITY PRESS
NEW YORK

Columbia University Press
Publishers Since 1893
New York   Chichester, West Sussex
cup.columbia.edu

Library of Congress Cataloging-in-Publication Data
Names: Sykes, L. R., author.
Title: Silencing the bomb : one scientist's quest to halt nuclear testing /
Lynn R. Sykes.
Description: New York : Columbia University Press, [2017] |
Includes bibliographical references and index.
Identifiers: LCCN 2017010287 (print) | LCCN 2017044883 (ebook) |
ISBN 9780231544191 (electronic) | ISBN 9780231182485 (cloth : alk. paper)
Subjects: LCSH: Nuclear weapons—Testing. | Comprehensive Nuclear-Test-Ban
Treaty (1996 September 10) | Nuclear arms control. | Sykes, L. R.
Classification: LCC U264 (ebook) | LCC U264 .S96 2017 (print) |
DDC 341.7/34—dc23
LC record available at https://lccn.loc.gov/2017010287

Columbia University Press books are printed on permanent
and durable acid-free paper.

Printed in the United States of America

Cover design: Milenda Nan Ok Lee

Cover image: Corbis Historical Collection / © Getty Images

CONTENTS

ACKNOWLEDGMENTS

I thank my wife, Kathleen M. Sykes, for her careful editing and advice. I thank Dan Davis, Paul Richards, and Frank von Hippel for carefully reading the manuscript and for their comments. I have worked with each of them on nuclear arms control issues for many decades. I also appreciate my many interactions on test ban issues with Charles Archambeau, Ola Dahlman, Göran Ekström, Jack Evernden, David Hafemeister, W-Y Kim, Peter Marshall, Meredith Nettles, and Gregory van der Vink. I also thank Kevin Krajick for his editing.

Many nongovernmental organizations (NGOs) in the United States have worked for decades to achieve nuclear test ban treaties. They include the following:

Arms Control Association
Center for Defense Information
Council for a Livable World
Federation of American Scientists
Friends Committee on National Legislation
Friends of the Earth
Institute for Science and International Security
Lawyers Alliance for World Security
Natural Resources Defense Council
Peace Action
Physicians for Social Responsibility
Stockholm International Peace Research Institute

I thank all of them for their long contributions to nuclear test bans and arms control.

INTRODUCTION

I have long wanted to write about my fifty years of work toward bringing about an international treaty that would completely ban the testing of nuclear weapons. I was fortunate to have participated in the negotiation of a nuclear test ban treaty with the Soviet Union in 1974. For more than fifty years I have also witnessed firsthand strenuous opposition in the United States to test bans by a cast of characters who sometimes acted from nefarious motives. The use of nuclear weapons is an issue that threatens our very existence. This is a story I believe needs to be told.

The quest for a complete ban on nuclear testing is now more than sixty-five years old. The process of ratification could exceed the seventy-two years it took from 1848 until 1920 to enact a constitutional amendment guaranteeing women the right to vote in the United States. The treaty has yet to enter into force because all countries possessing either nuclear weapons or reactors have not ratified it. Nevertheless, Russia, the United States, China, Britain, France, and Israel have not tested nuclear weapons since they signed it in 1996; India and Pakistan, which did not sign the treaty, have not tested since 1998. In these important ways the treaty has been very successful.

A major nuclear exchange between the superpowers would be a disaster of unprecedented destruction and horror, the worst in all of human history. An exchange between India and Pakistan alone could kill more than half a billion people and affect other countries, particularly China. Although the Cold War is over, weapons of the superpowers are on hair-trigger alert and could be fired in response to false alarms or by unauthorized users.

Other tasks pulled at me through the years: research on earthquakes in the greater New York City region, revisiting early work on plate tectonics, and long-term earthquake prediction as new information became available, as well as working as a consultant to New York State regarding the likelihood of earthquakes near nuclear power plants located along the Hudson River not far from New York City. During my forty years as a professor prior to my retirement in 2005, I advised about thirty graduate students at Columbia University, raised funds for their support, and usually taught two classes each year in a dozen different areas of the earth sciences, environmental hazards, and the nuclear arms race.

Now, at eighty years old, I have made time to reflect upon my life—professionally and personally. My undergraduate years at the Massachusetts Institute of Technology opened doors to me, both scientifically and culturally. The Lamont Geological Observatory of Columbia University, where I landed as a graduate student in 1960 after college with a degree in geology and geophysics, had been formed only a dozen years earlier. I was in the right place at the right time as I became involved in the birth of plate tectonics and the development of methods to verify a Comprehensive Nuclear Test Ban Treaty. My chosen field of seismology, the study of earthquakes, is the primary science and technology for detecting, locating, and identifying underground nuclear tests. Methods for examining earthquakes are very similar to those for nuclear explosions.

In 1966, after halting research on another project, I worked on the mechanisms of earthquakes and demonstrated that new seafloor was being formed along mid-oceanic ridges and that continental drift, long rejected by many scientists, was a reality. I was fortunate that the early part of my career and the following decade coincided with the Golden Age of funding of the earth sciences in the United States.

My years as a graduate student, from 1960 to 1965, were a particularly frightening time during the nuclear arms race. Intercontinental ballistic missiles were being deployed, and very large nuclear devices and weapons were being tested in the atmosphere. The Cuban Missile Crisis of 1962 brought the world alarmingly close to nuclear war. Deeply concerned, it was at this time that I became committed to do whatever I could in forwarding the process of better verification of nuclear tests. Detection and identification of underground explosions have been two of the prime concerns of those involved with a full test ban treaty.

My decision to specialize in studying earthquakes brought with it an awareness of the importance of a full halt to the testing of nuclear weapons as well as further steps toward nuclear arms control. Realizing the devastating consequences of a nuclear conflict, I made a major commitment to do whatever I could to affect the signing and ratification of a comprehensive test ban that would encompass the monitoring of then difficult-to-identify underground explosions and bring an end to atomic testing.

My work on nuclear tests and their detection and identification began in 1965, soon after I completed graduate work at the Lamont Geological Observatory. Most of the following chapters are devoted to work I did personally and to my interaction with key individuals, government agencies, and other organizations for the next fifty years. I have emphasized contributions made through the years to better identify nuclear explosions; bureaucratic struggles in the United States over nuclear monitoring, including claims that Russia cheated on two test ban treaties; the desirability of a full test ban treaty; other steps toward the control of nuclear weapons; and the dangers of nuclear war. I have attempted to the best of my ability to cover both the political and the technical and scientific aspects of the long quest for a complete test ban.

My involvement with negotiations for the Threshold Test Ban Treaty, which took me to Russia in 1974, led me to work for the next fifteen years on estimating the sizes (yields) of Soviet nuclear explosions. I have been involved in long debates in the U.S. government, sometimes referred to as the "yield wars," where some agencies and people in the United States claimed that the Soviet Union was cheating on the Threshold Treaty. Several times during the 1980s, I was called before Congress to testify concerning the sizes of Russian explosions and was able to demonstrate that the USSR, in fact, was not cheating, as some governmental hawks would have liked us to believe. I argued further that the United States and other countries needed to move toward a complete and verifiable ban on nuclear testing.

My work in the 1990s also involved dealing with the possibilities of evasive nuclear testing, a subject that was of paramount concern and a stumbling block to a full test ban in the United States in 1963 and again in 1999.

I have tried to make this book accessible to a wide audience of educated people, including students and others interested in learning more about the development of the nuclear age through the eyes of an insider, as well as those involved with arms control, public policy, and the history of science. It is not intended as a textbook or a journal article on nuclear tests, monitoring, or earthquakes.

I decided to begin this book with my sudden trip to Moscow in 1974 to participate in the negotiation of a treaty to limit very large underground nuclear explosions. I go on to describe the development of nuclear weapons and early attempts to ban their testing. While the book is in historical order, I discuss major scientific and political developments as they occurred within the context of a particular subject. For example, in chapter 4 I emphasize claims in the United States in the early 1960s that other countries, particularly Russia, could test evasively and thus escape detection. I then return to evasion in chapter 15 during the Senate's rejection of the full test ban treaty in 1999. In chapters 5, 9, and 10 I describe long continuing arguments about determining the sizes or yields of Soviet nuclear explosions, their overestimation by the United States, and the final resolution of the "yield wars."

One of the major challenges in monitoring a full test ban has been differentiating the seismic signals of underground tests from those of the many earthquakes that occur every day. In chapters 6, 7, 11–14, and 16 I enumerate major scientific improvements over time in identifying underground tests.

Peaceful nuclear explosions (PNEs), which are covered in chapter 8, are relevant to a full test ban treaty because they cannot be distinguished from tests of weapons. The Soviet Union conducted many PNEs in thick salt deposits, where it is possible to construct large underground cavities and then detonate small muffled explosions evasively in them. In chapter 13 I describe the successful negotiation of the Comprehensive Nuclear Test Ban Treaty and of an extensive set of global stations set up to monitor it. Chapter 16 includes reports by the U.S. National Academies of Science, Engineering, and Medicine, in which I participated, that focused on the monitoring of that treaty and on ongoing efforts to insure that the U.S.

stockpile of nuclear weapons will continue to work into the future. I close with a final chapter on the dangers of nuclear war, the control of nuclear weapons, limiting their delivery systems, and possible ways forward.

I am currently at work on a companion book, *Plate Tectonics and Great Earthquakes: One Scientist's Perspective on Fifty Years of Earth Shaking Events*, which covers my involvement in groundbreaking discoveries in the development and testing of plate tectonics during the 1960s, as well as later work on great earthquakes, long-term earthquake prediction, shocks within the North American plate, risks to nuclear power reactors, and much more. It includes two chapters on my personal life and education.

The intertwining of science with arguments about halting nuclear testing continues today as it has since the early 1950s, when the first attempts to stop tests occurred. A number of heroes and villains stand out in the long road to a total ban on nuclear testing. Here is my story.

SILENCING THE BOMB

1

A HURRIED TRIP TO MOSCOW IN 1974 TO NEGOTIATE THE THRESHOLD NUCLEAR TEST BAN TREATY

My phone rang, waking me from a sound sleep at 5:00 a.m. Only a few hours before, I had returned home to New York City from three days of canoeing in the Adirondacks during the Memorial Day weekend of 1974. The caller was Eric Willis of the Department of Defense, who asked me to join a U.S. team that was leaving that evening for Moscow to negotiate what was to be called the Threshold Test Ban Treaty, or TTBT.

As I struggled to wake up, I agreed to be part of the group. Eric thanked me and told me I needed to get down to Washington, DC, immediately to pick up my visa before the Soviet embassy closed at noon. I took an 8:00 a.m. flight and, in my rush to get there, did not have time to put together clothes, scientific materials, or money. Hours later, I returned to New York, hurriedly packed, phoned a few people to let them know what I was doing, went to the bank, and headed back to Washington on an Eastern Airlines shuttle.

I had only the tail number of the Air Force plane that was leaving from Andrews Air Force Base with the U.S. delegation. My Eastern Airlines pilot radioed ahead that I was coming, and the plane to Moscow waited for my slightly late and breathless arrival. I had no idea if our sojourn in Moscow would last a few days or several weeks. Neither did the other members of the delegation. We ended up spending close to a month in Moscow.

Our group was initially described as one for "technical discussions" of a possible Threshold Treaty. A staff member for Henry Kissinger, President Nixon's secretary of state, informed us that our purpose was to explore whether the Russians were sincere about negotiating a bilateral treaty with the United States to limit the size of underground nuclear explosions. We soon learned that they indeed were interested.

In an effort to counter the ongoing political storm over Watergate, Kissinger had advised the U.S. government to seek a relatively quick and modest arms control treaty with the Soviet Union. Negotiating a more complex and comprehensive treaty, such as a Strategic Arms Limitation Treaty (SALT) to limit long-range delivery systems for nuclear weapons, would have been a much too ambitious and lengthy endeavor. The Soviet Union was attracting international attention in 1974 by pressing for a Comprehensive Nuclear Test Ban Treaty (CTBT) while the United States resisted, citing concerns about verification and the reliability of the nuclear stockpile. At a meeting between Kissinger and Soviet foreign minister Andrei Gromyko in March 1974, they agreed in principle to seek a ban on nuclear explosions above some yet-to-be-determined size. I did not know this when I joined the negotiating team on the plane to Moscow.

Our U.S. Air Force plane to Moscow, which had no passenger windows, stopped in Copenhagen. NATO allies were then informed about the general purpose of our mission. When we arrived in Moscow, Soviet "handlers" met us and took us to the Rossiya Hotel. Adjacent to Red Square, its twenty-one-story tower then loomed above the Kremlin walls and the cupolas of Saint Basil's Cathedral. Our "handlers" and others from the KGB wore expensive sport coats or suits that definitely were not available to ordinary Soviet citizens.

I knew the Rossiya from stays in Moscow for an international meeting on geophysics in 1971 and as part of a U.S. delegation on earthquake prediction in 1973. Each floor had two kindly looking elderly women who observed our comings and goings. Except for Walter Stoessel, the U.S. ambassador to the Soviet Union, a State Department employee, and interpreters, no one else in our delegation knew Russian. Most had not been to the Soviet Union previously. From my one-semester course in Scientific Russian at Columbia, I knew the Russian (Cyrillic) alphabet and some Russian words. I could not hold a conversation in Russian, but I was able to read signs, find my way around central Moscow, travel on the subways, and order breakfast from coffee bars in the Rossiya.

Our delegation included representatives from the Department of Energy, which funded the laboratories that developed and tested nuclear weapons, the departments of Defense and State, the Arms Control and Disarmament Agency (ACDA), and the Joint Chiefs of Staff. Ambassador Stoessel, a career diplomat, headed our delegation. Norman Terrell, who worked for Helmut Sonnenfeldt, senior counselor to Kissinger, made many important decisions. Terrell was the primary link to Kissinger and a panel called the "back-stopping" committee, a group of experts on nuclear testing, back in Washington.

Picking members of the negotiating team that was sent to Moscow had to have been a major concern. It is interesting to see who was picked for the U.S. team and who was not. Seismology was the main technology used to monitor the sizes of explosions and hence the treaty's threshold. Of the twelve members of the team, three of us were seismologists. Our delegation was unusual in that it contained two university seismologists, Eugene Herrin from Southern Methodist University in Dallas and me. Carl Romney, a chief scientist at the Advanced Research Projects Agency (ARPA) in the Pentagon, had formerly worked at the Air Force Technical Application Center (AFTAC).

It is strange that no one from AFTAC, the agency responsible for operating the U.S. classified Atomic Energy Detecting System (AEDS), was a member of the negotiating team. Many U.S. government officials probably considered that Romney and Herrin had the greatest expertise in understanding the U.S. classified capabilities in seismology. Seismologists from the U.S. weapons labs were not part of the delegation either. Donald Springer of the Livermore Lab would have been an excellent member, but he was not chosen.

Warren Heckrotte of Livermore, who was familiar with previous test ban negotiations, was replaced by Michael May, a high-level administrator at Livermore and an excellent physicist, about halfway through the negotiations when officials in Washington concluded that the Soviets were serious about negotiating a treaty. General Edward Rowny of the office of the U.S. Joint Chiefs of Staff, who was responsible for negotiations with the Soviet Union for the Strategic Arms Limitation Talks (SALT), also replaced one person from the Defense Department. Eric Willis of ARPA, a geochemist familiar with measuring radioactive isotopes, remained a member.

I had worked in geophysics at the Lamont Geological Observatory since 1960. My main expertise in 1974 was studying long-period seismic surface waves, plate tectonics, and the discrimination of the signals of underground explosions from those of earthquakes.

As it turned out, Herrin and I came to form a buffer between two warring factions within U.S. government officials at the negotiations—those who sought a verifiable treaty and hawks who opposed it, seeing it as an unadvised step toward a complete ban on nuclear testing. One member of the delegation from the Defense Department, a former professor of engineering at Columbia University, described himself to me as a "professional bastard." Fortunately, he did not have much impact on the negotiations.

After two weeks, some agencies in Washington wanted to replace Herrin and me with others from within the government. Ambassador Stoessel successfully resisted replacing us on the grounds that we made significant contributions to the technical discussions and to the balance of the delegation.

About two weeks into the negotiations, the ambassador asked each of us in the delegation to state whether the allowed explosive yields of nuclear tests should be set at a threshold of 100 or 150 kilotons (a kiloton, abbreviated kt, is equivalent in explosive energy to one thousand tons of the chemical explosive TNT). Officials in Washington had sent those numbers to the ambassador. Herrin and I said that seismic waves generated by explosions of either of those yields could be detected readily all over the world, and their seismic signals could be distinguished easily from those generated by earthquakes.

Our delegation had to be careful about classified materials. We were instructed not to leave any in our hotel rooms or on the conference room table, where they could be photographed by cameras hidden in the ornate ceiling of our meeting room in Moscow. We were not to talk about classified materials or the negotiations in Soviet cars that transported us to and from meetings. We should not talk to any of our scientific colleagues in Moscow who were not members of the Soviet delegation. I went to an opera in the Kremlin one evening where by chance I met Malcolm McKenna, a famous U.S. expert on paleontology, who held joint appointments at the American Museum of Natural History and our department at Columbia. He was en route to Mongolia to negotiate the reopening of joint scientific work between U.S. and Mongolian paleontologists at a famous dinosaur area that had not been visited by Western scientists since the 1930s. I could not tell him why I was in Moscow other than it involved earthquakes and explosions.

A general in our delegation planned to attend the opera and asked one of the "handlers" what he should do if he got lost. The answer, in very accented English, was "Do not worry." Obviously there was no chance of that because we were under close surveillance. One weekend our delegation traveled overnight by train to Leningrad for sightseeing. Upon our arrival at the train station, military officers saluted one of the Soviets accompanying us. Although she was not a scientist or a member of the Soviet delegation, she was obviously a high-ranking official.

Our delegation met formally with representatives from the Soviet Union about four times a week, when members made formal presentations on specific topics followed by questions from the other delegation. I presented one on the determination of seismic magnitudes. At other times we met in the U.S. embassy, where we went over in detail the papers that were to be presented to the Soviets. Some of us walked from the embassy to our hotel for exercise and fresh air. One of our Soviet handlers later remarked that we were fast walkers.

The Soviet delegation contained at least two members who worked on peaceful nuclear explosions (PNEs). They stated that PNEs were greatly needed for their national economy such as constructing a major canal. We met informally for dinner and drinks a few times and traveled with members of their delegation on weekends. At other times we ate with members of our delegation at our hotel. (Interestingly, we received better food after our ambassador formally announced agreements on specific topics, such as using yield as a measure of the threshold.) Roland Timerbaev of the Soviet Ministry of Foreign Affairs took two of us to the nearby historic town of Zagorsk another weekend. I was surprised that he, a Muslim, said he expected the Soviet Union to eventually accept the Russian Orthodox religion, which it did many years later.

Kissinger and Nixon negotiated the final details of the TTBT in the first few days of July 1974. A few days later, Nixon and Soviet premier Leonid Brezhnev signed the treaty, which set the underground testing threshold at the more conservative 150-kiloton level. I flew home from Moscow to Frankfurt on a U.S. Air Force transport plane that had delivered a red convertible to Moscow, Nixon's gift to Brezhnev, who was known to like fast cars.

I think the Threshold Test Ban Treaty brought us another step closer to a Comprehensive Nuclear Test Ban Treaty (CTBT) by outlawing all nuclear weapons tests above the yield threshold. Nevertheless, I had no idea that twenty-two years would pass before a full test ban (a CTBT) finally would be approved by the General Assembly of the United Nations and signed by President Clinton and leaders of many other nations in 1996.

As of mid-2017, 183 out of a total of 196 states had signed the Comprehensive Test Ban Treaty (CTBT), and 166 had both signed and ratified it. Those signing in 1996 included all states that acknowledged having nuclear weapons. The 1974 TTBT was signed about midway between the adoption of the CTBT in 1996 and the first serious calls for a Comprehensive Test Ban Treaty in 1954, soon after the United States and the Soviet Union had tested very large hydrogen (thermonuclear) weapons in the atmosphere.

Nevertheless, although the major nuclear countries signed the CTBT in 1996, it can enter into force only after it has been signed and ratified by all forty-four countries that have either nuclear weapons or reactors. On-site inspections become possible under the treaty after its entry into force. Russia, Britain, and France have signed and ratified the treaty.

Entry into force unfortunately remains an elusive goal. Neither the United States nor China nor Israel has ratified the treaty, even though each signed it in 1996. India, North Korea, and Pakistan, which now possess nuclear weapons, have not signed the treaty. Though not ratified, the CTBT has been very successful in that the countries that signed it have not tested nuclear weapons of military significance since 1996. India and Pakistan have not tested since 1998.

Nuclear testing in the atmosphere, in space, and underwater was outlawed by the Limited Test Ban Treaty (LTBT) of 1963. Nevertheless, the LTBT neither prohibited underground testing nor stopped the continued development of new nuclear weapons, as the 1996 CTBT does. While the LTBT did prevent great amounts of radioactive debris from entering the atmosphere and oceans, it was largely a public health measure.

2

DEVELOPMENT AND TESTING OF NUCLEAR WEAPONS

DEVELOPMENT OF NUCLEAR WEAPONS

During World War II, the United States government created the Manhattan District, often called the Manhattan Project, to develop the first nuclear weapons. It brought together leading scientists to work on nuclear weapons and organized facilities for producing their nuclear ingredients. Physicist J. Robert Oppenheimer, a professor at the University of California at Berkeley, was chosen as the director of the Los Alamos Laboratory, which designed the first nuclear weapons.

The United States, fearing that Nazi Germany would get nuclear weapons first, constructed huge facilities to obtain two types of materials that can produce nuclear explosions by splitting (fissioning) the nucleus of certain atoms. The nucleus of an atom, consisting of protons and neutrons, accounts for most of the weight of an atom. The energy liberated in nuclear fission is huge. It is much greater than that in ordinary chemical reactions, which merely involve the lightweight electrons that surround the nucleus.

One facility at Oak Ridge, Tennessee, was constructed to separate a rare form of the element uranium, abbreviated U 235, from the more abundant form of uranium, U 238. A factory at Hanford, Washington, produced plutonium in nuclear reactors. Plutonium (abbreviated Pu) is a man-made element that is not present naturally on Earth today.

Most elements consist of different isotopes (or flavors) that differ only very slightly from one another. Consequently, they are difficult to separate. For example, all of the various isotopes of uranium consist of

92 protons and 92 electrons. The isotope U 235 of uranium contains 235 – 92 = 143 neutrons, whereas U 238 consists of 238 – 92 = 146 neutrons.

In 1938 Otto Hahn, Fritz Strassmann, and Lise Meitner of the Kaiser Wilhelm Institute for Chemistry in Berlin, bombarded uranium atoms with neutrons and discovered that they could be fissioned, or broken into much lighter elements. Meitner, who was Jewish, fled to the Netherlands with the help of Hahn in July 1938 and went on to Sweden. In November 1938 Hahn discussed the results of his ongoing experiments with Meitner and Danish physicist Niels Bohr. Meitner and her nephew Otto Robert Frisch worked out the basic mathematics of nuclear fission in Sweden; it was Frisch who called the process nuclear fission. They realized that mass, m, was converted into a vast amount of energy, E, by Einstein's famous equivalence of energy and mass, $E = mc^2$ where c is the speed of light.

In a second paper on the fissioning of uranium in February 1939, Hahn and Strassmann predicted the liberation of neutrons during the fission process. A chain reaction involving the continued fission of atoms was central to the liberation of immense amounts of energy and the development of nuclear weapons. A chain reaction occurs when a neutron causes an atom of U 235 to fission and to produce more neutrons, which go on to fission additional uranium atoms. Hahn and Strassmann, who did not leave Germany, received the Nobel Prize in 1944 for their discovery. Meitner should have been included.

Fission involves the breakdown of either the plutonium isotope Pu 239 or uranium U 235 into lighter elements, typically ones near the middle of the periodic table of elements, and the release of huge amounts of energy as nuclear mass is converted into energy. In contrast, chemical reactions involve the release of much smaller amounts of energy per pound (or kilogram) and do not involve the change of one element into others as in nuclear reactions.

During World War II, the United States quickly developed two types of nuclear weapons. An explosive device called the *Gadget* with a yield of 21 kilotons was tested in New Mexico on July 16, 1945. (A kiloton, or kt, is the equivalent energy released by 1000 tons of TNT.) *Gadget* consisted of 13.5 pounds (6.1 kg) of plutonium and about 5000 pounds (2270 kg) of high explosives to compress the plutonium into a denser mass. Ten seismic stations at distances of 270 to 700 miles (435 to 1130 km) recorded the explosion. An untested U 235 bomb called *Little Boy* was exploded over Hiroshima, Japan, on August 6, 1945. It weighed about 8000 pounds (3630 kg) and had a yield of about 13 kilotons.

A plutonium nuclear weapon called *Fat Man* was detonated over Nagasaki, Japan, on August 8, 1945 with the same yield as that of the New Mexico device. The two nuclear weapons helped to bring the Pacific war to an end and ushered in the atomic age. A large bomber could carry one of the nuclear weapons of 1945 vintage, but these bombs were extremely heavy. The U 235 weapon used against Hiroshima, though not tested in a nuclear explosion beforehand, was developed by a group of some of the best scientists from many nations and to exacting tolerances. It would be difficult even today for a nonnuclear state to count on a similar device working without being fully tested, and it would be too heavy to place on a relatively crude missile.

The United States government initiated a classified program called Long-Range Detection in 1947, under the direction of the Air Force, to monitor nuclear testing by other counties, especially the Soviet Union and later China. The classified monitoring network, called the Atomic Energy Detection System (AEDS), was, and still is, operated by the Air Force Technical Applications Center (AFTAC). Publications by AFTAC in 1997 and in 2009 by seismologist Carl Romney, who worked for them for many years, describe this early work on monitoring in more detail. In 1948 the United Kingdom and Canada were involved in monitoring nuclear tests set off by the United States on towers at Eniwetok Atoll in the western Pacific. Those explosions were detected by sampling airborne radioactivity but were not picked up by seismic stations at distances greater than 500 miles (800 km).

The Air Weather Service of the Air Force began flying military aircraft between Alaska and Japan with special filters to try to detect radioactive debris carried by winds from possible atmospheric nuclear explosions by the USSR. Radioactive debris was captured by one such flight on September 3, 1949, and by one from Guam to Japan two days later. U.S. experts on radioactive isotopes identified collected debris as having been generated by a Soviet explosion of a plutonium device sometime between

August 26 and 29, 1949. Scientists used meteorological observations to backtrack its location to somewhere in Central Asia. President Truman announced to the public on September 23 that the Soviet Union had conducted a nuclear test.

When Premier Joseph Stalin learned about the Hiroshima explosion, he ordered the rapid development of atomic weapons by the USSR. Soon after World War II, the United States proposed the international control of fissionable materials and a halt to the arms race, but Stalin was not responsive. It was decades before it became known publicly that the Soviet explosion of 1949 was an exact copy of the U.S. *Fat Man* weapon of 1945.

British scientist Klaus Fuchs, who was present at Los Alamos during the Manhattan Project, obtained its design by espionage. He had been a member of the Communist Party in Germany before fleeing to England in the early 1930s. Returning to Britain after World War II, Fuchs confessed he was a spy in 1950 and was convicted. He served nine years in prison and then immigrated to East Germany, where he died in 1988. The USSR built a reactor in the Ural Mountains to obtain plutonium for the 1949 explosion.

The Soviet test of 1949 led Truman to order the rushed development of larger fission as well as thermonuclear weapons. The latter are often called "The Super" hydrogen or fusion weapons. The fusion of hydrogen takes place only at exceeding high temperatures, millions of degrees, like those in the sun. Stars liberate vast amounts of energy primarily by converting hydrogen by fusion into helium. Fusion weapons involve the conversion of one or two of the heavy hydrogen isotopes, deuterium and tritium, into helium. Fusion converts mass into huge amounts of liberated energy.

The United States made several improvements in the yield-to-weight ratio of fission weapons. In a key development called "boosting," heavy isotopes of hydrogen—deuterium and/or tritium—were used to increase the number of neutrons bombarding the fissile material in the core of a fission device. Boosting permitted the weight of a weapon to be reduced considerably. The United States tested a boosted device named *Item* with a yield of about 45 kilotons in May 1951.

In the *Mike* nuclear explosion of 1952 (figure 2.1), the United States tested the concept of igniting a full-scale *fusion* explosion with a small

FIGURE 2.1

The ***Mike*** thermonuclear (hydrogen bomb) explosion.

Photo by U.S. Atomic Energy Commission.

fission explosion, called either an initiator, a primary, or a trigger. A roomful of heavy equipment was needed to maintain its thermonuclear fuel, deuterium, at very low temperatures. Its yield of 10.4 megatons (Mt), or 10,400 kilotons, was about 800 times the yield of the Hiroshima explosion of 1945. *Mike* obliterated part of the Pacific island of Elugelab in the Marshall Islands. The arms race soon accelerated, and nuclear explosions of very high yield were developed and then deployed as weapons by the U.S. Air Force.

The United States went on to develop fusion weapons with a solid fuel called lithium deuteride (the lightweight element lithium combined with one of the heavy isotopes of hydrogen called deuterium). That fuel is stable at room temperatures, meaning it could be used and deployed for fusion weapons. The United States tested this device first at Bikini Atoll in the *Bravo* explosion of March 1, 1954. Its yield of 15,000 kilotons (15 Mt) was much larger than expected, because the contribution of an isotope of lithium, Li 7, to fusion reactions was not foreseen. During the 1950s, the United States went on to develop and test a number of thermonuclear weapons in the megaton range.

Bravo produced large amounts of radioactive fallout, of which the nearby Japanese fishing boat *Lucky Dragon* received high levels. Twenty-three of its crew developed radiation sickness by the time the boat docked in Japan. Its captain died of leukemia six months later. Indian prime minister Jawaharlal Nehru made the first proposal that year for a halt to and a ban on nuclear testing.

The Soviets had not sat idle. They detonated two fission explosions in 1951 and their first thermonuclear device, with a yield of 400 kilotons, on August 12, 1953. Seismic stations in the United States and abroad recorded the test. In 1955 the USSR detonated a weaponized version of the 1953 device, with a yield of 215 kilotons, and their first thermonuclear device with a yield exceeding one megaton.

In 1957 and 1958 the Soviet Union and the United States conducted many thermonuclear tests, a number them in the megaton range. In 1952 the United Kingdom tested its first fission device, and in 1957 its first hydrogen bomb.

A megaton nuclear explosion would cause damage and deaths over a huge area from thermal radiation (heat), the atmosphere blast wave, and high-energy radiation, as illustrated in figure 2.2 for the New York City area.

SUMMARY OF NUCLEAR WEAPONS TESTS BY VARIOUS COUNTRIES AND LONG-RANGE (STRATEGIC) NUCLEAR WEAPONS OF RUSSIA AND THE UNITED STATES

Without testing, a country cannot be certain that a new nuclear weapon will work. Of course, many weapons would not have entered the stockpiles of various nations if a full ban on testing had been enacted in 1963, not when it was finally signed in 1996.

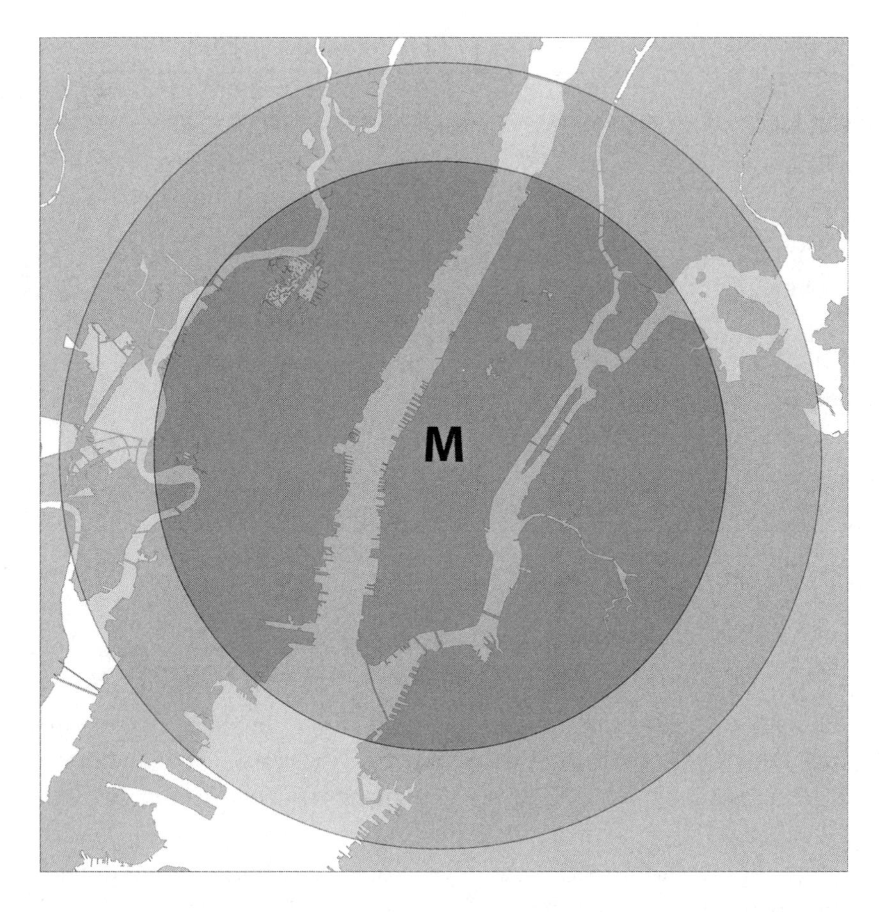

FIGURE 2.2

Nuclear firestorm created by the detonation of an 800-kiloton weapon above New York City. *M* denotes midtown Manhattan. No one survives in the central fire zone (area 90 square miles, or 23,000 hectares) that includes most of the island of Manhattan; a firestorm is likely in the larger shaded area.

Computations courtesy of Theodore Postol of MIT, 2016.

Table 2.1 lists the first fission test, the first fusion (hydrogen bomb) explosion, and the last underground test by the various nuclear powers as of late 2016. They have now conducted a total of more 2000 nuclear tests.

TABLE 2.1 Number of Nuclear Tests and Dates as of September 2016

COUNTRY	FIRST FISSION TEST	FIRST FUSION TEST	MOST RECENT TEST	TOTAL TESTS
United States	1945	1952	September 1992	1030*
Russia	1949	1953	October 1990	715*
United Kingdom	1952	1958	October 1991	45**
France	1960	1968	January 1996	210
China	1964	1967	July 1996	45
India	1974		May 1998	4–6
Pakistan	May 1998		May 1998	2–6
Israel & South Africa?	September 1979?		September 1979?	
North Korea	October 2006		September 2016	5

*Some tests involved two or more explosions close in time and location.
** Twenty-four were joint tests with the United States.

In their 2009 book *The Nuclear Express*, Thomas Reed and Danny Stillman argue that many of the countries that acquired nuclear weapons did so with the help of other nations. Nuclear weapons designers Reed and Stillman worked at the Livermore and Los Alamos weapons laboratories. Scientists from the United Kingdom helped develop the U.S. atomic bombs at Los Alamos and hence knew much about their design. The Soviet Union acquired the design of its first device from the United States by espionage. China initially received help from the USSR on its atomic bomb before Russia cut off aid over fear of what Mao might do with atomic weapons.

Reed and Stillman state that Britain aided France with the design of its hydrogen bomb; France helped Israel with its atomic weapons program. India obtained plutonium for its 1974 nuclear test from a reactor that Canada and the United States had supplied it under the Atoms for Peace program. They also conclude that the first Pakistani atomic device likely was tested at China's Lop Nor site in 1990, but others disagree with their claim. Pakistani engineer A. Q. Khan transferred uranium enrichment technology to Libya, Iran, North Korea, and perhaps other states.

TABLE 2.2 Numbers of Nuclear Weapons of Various Countries as of 2013

COUNTRY	DEPLOYED	OTHER	TOTAL 2012	TOTAL 2013
United States	2150	5550	8000	7700
Russia	1800	6700	10,000	8500
United Kingdom	160	65	225	225
France	290	10	300	300
China		250	240	250
India		90–110	80–100	90–110
Pakistan		100–120	90–110	100–120
Israel		80	80	80
North Korea			2	3
Total*	4400	12,865	19,000	17,265

*Totals from Stockholm International Peace Research Institute Yearbook, 2013 (http://www.sipriyearbook.org).

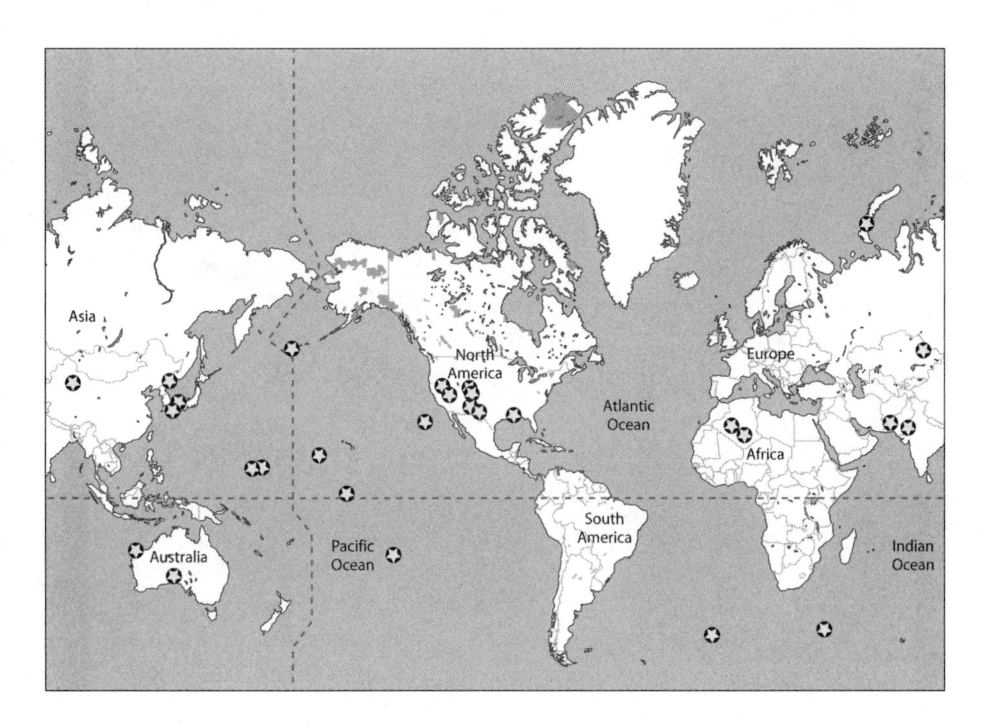

FIGURE 2.3

Areas where nuclear explosives have been tested. The figure also includes the Hiroshima and Nagasaki explosions of 1945. Sites of Soviet and U.S. peaceful nuclear explosions are not shown. Whether a nuclear explosion was tested at the site shown in the southwestern Indian Ocean is discussed later.

Source: atomicarchive.com.

Table 2.2 lists the numbers of nuclear warheads of nine countries. Those that can be delivered at a moment's notice are classified separately from those ("other") held in reserve or being dismantled. Note that Russia and the United States have by far the greatest numbers of nuclear weapons. Figure 2.3 shows the sites of nuclear weapons tests.

3

FROM THE EARLY NEGOTIATIONS TO HALT NUCLEAR TESTING TO THE LIMITED TEST BAN TREATY OF 1963

I highlight here some of the proposals, negotiations, and problems in working toward either a limited (LTBT) or a full comprehensive nuclear test ban (CTBT) during the administrations of Dwight D. Eisenhower and John F. Kennedy.

Following a call by India's Prime Minister Nehru in 1954 for a pause in nuclear testing, several other countries presented proposals for a halt in testing. Formal negotiations toward a test ban began in the second half of the 1950s and continued into the early 1960s.

PROS AND CONS OF BANNING NUCLEAR TESTS

Debate has raged for more than sixty years over whether a Comprehensive Nuclear Test Ban Treaty (CTBT) is in the interest of the United States. I touch here briefly on some of the main reasons a CTBT has been either proposed or opposed.

It should be remembered that a CTBT involves a ban on the testing of nuclear weapons, not a ban on their delivery systems such as aircraft and missiles. It does not prohibit the manufacture of additional weapons that have already been tested. While a CTBT is an important step, it is but one component of an overall effort to control nuclear arms. A CTBT may help to promote other arms control initiatives and agreements.

Many of the arguments for and against a CTBT (see Table 3.1) have been enunciated for decades. Some are new with the 1996 treaty. It is

TABLE 3.1 Pros and Cons of Banning Nuclear Tests

PRO	CON
1. The U.S. lead in nuclear weapons will diminish without a CTBT.	Others will cheat and erode the U.S. lead. National security is best enhanced by continued testing of new weapons.
2. The treaty will slow or halt the nuclear arms race (stop continued development of weapons by nuclear powers; stop new states from testing and acquiring nuclear arms).	This is wishful thinking.
3. The treaty is verifiable down to explosions of very tiny yields, which are not of military significance.	Even an explosion between zero and a few hundred tons' yield gives an advantage to others. Determined cheaters will test important weapons evasively below verification limits.
4. The treaty is for zero nuclear yield.	The lower yield limit is vague; it should be a low yield threshold treaty.
5. Russia and China are unlikely to be able to deploy a new long-range nuclear weapon without several multikiloton tests, which would be detectable even with evasive testing.	Tactical and other weapons with yields below one kiloton are important (x-ray laser, electromagnetic pulse warhead).
6. The Stockpile Stewardship program is more effective than originally thought.	The program is not good enough without the ability to test existing and new weapons.
7. U.S. classified verification capabilities are better than the International Monitoring System (IMS). The United States makes its own determinants for a suspicious event.	The IMS is not good enough. The United States will surrender assessment of an event as being nuclear to the IMS and the UN.
8. The United States can pursue a few arms control agreements simultaneously.	Strategic arms limitations are more important than test bans.
9. Tests of thermonuclear weapons are unlikely to be concealable even with evasive testing. This would end development of advanced types of weapons.	Weapons labs need testing to retain top scientists.
10. If problems occur with existing weapons, the United States can withdraw from the treaty within six months.	To prevent deterioration of weapons, nuclear testing is necessary.

TABLE 3.1 *(Continued)*

PRO	CON
11. On-site inspections are possible once the treaty enters into force. The UN General Assembly could bring it into force without rogue states once the United States and China ratify.	Entry into force will never occur because forty-four countries must ratify the CTBT.
12. Components of nuclear explosives can be remanufactured and old components replaced.	Some components of nuclear explosives cannot be remanufactured.
13. It is illusory to believe that amendments to the treaty, such as making it of limited duration, can be negotiated with other signatories. The U.S. Senate should enact U.S. safeguards.	The treaty should be amended to make it of limited duration.
14. The treaty helps to establish the norm that the only value of nuclear weapons is for deterring their use by others.	Other countries and terrorists may not accept this view. Rogue nations will test. Nuclear weapons are needed to attack chemical and biological weapons.
15. U.S policy is that it does not need new nuclear weapons.	The United States should develop and test new nuclear weapons; requirements change.
16. The treaty will strengthen international peace and security. International support for the treaty is very strong.	This is a false hope.
17. U.S., Russian, and Chinese weapons are already one-point safe; hence, they do not need tiny nuclear warheads.	To improve safety, add insensitive high explosives to submarine warheads.

important to note that verification capabilities have increased, especially in the past twenty years.

In addition to the statements listed in Table 3.1, several domestic arguments have been made for U.S. ratification of a complete test ban.

1. A CTBT is favored by a vast majority of Americans. Polls indicate that few people in the United States know that the Senate defeated the CTBT in 1999.

2. The United States would miss an historic opportunity to make the world safer for future generations.
3. Failure to ratify the CTBT would weaken the effectiveness of the Non-proliferation Treaty (NPT). Some nations may withdraw. (Thus far, however, only North Korea has withdrawn from the NPT.)
4. Failure to ratify would undercut the status of America as a world leader. Global standards do matter.
5. U.S. ratification may dissuade others such as India and Pakistan from conducting more tests. (They have not tested since 1998. Only North Korea has tested nuclear weapons in the twenty-first century.)

U.S. AND SOVIET PROPOSALS TO HALT TESTING

In July 1955 and again in 1956, Premier Nikolai Bulganin of the USSR proposed an end to nuclear testing. His plan, however, did not include verification. Understandably, the United States argued for verification of arms control agreements, especially those related to nuclear weapons. In 1955 the declared aim of the United States was to seek a comprehensive disarmament agreement that included verification.

The U.S. position on halting or limiting nuclear testing underwent several changes during the Eisenhower administration, which ran from 1953 through early 1961. In 1956 Adlai Stevenson, the Democratic nominee for president, suggested the United States might stop testing as a first step toward an agreement with the USSR. He reasoned that the United States could verify a ban on high-yield testing, ending the worst dangers of radioactive fallout.

The radioactive isotope strontium 90 produced by testing in the atmosphere was detected in human bones and teeth worldwide in the 1950s. Since the chemistry of strontium is similar to that of calcium, humans adsorb radioactive strontium (Sr 90) as if it were calcium in dairy and other foods. Sr 90 has a half-life of about twenty-eight years. (A half-life is the time it takes for a radioactive isotope to decay to half its original concentration.) Hence, it takes two hundred years for Sr 90 to decay to 1 percent of its initial value. In 1959 Lamont geochemist Larry Kulp and his colleagues, who made many of the measurements on human bones, published their results in the journal *Science*, drawing public and governmental attention to the issue of fallout from atmospheric testing.

By 1957 a number of scientists, public figures, and various organizations demanded an end to nuclear testing. Chemist and Nobel Laureate Linus Pauling of Cal Tech circulated a petition calling for a test ban; nine thousand scientists in forty-three countries signed it. In the United States, SANE, the Committee for a Sane Nuclear Policy, placed ads in major newspapers citing the perils of nuclear war. Thousands of letters protesting continued nuclear testing were sent to President Eisenhower. Many nongovernmental organizations in the United States and the UK who protested against continued testing became active at that time.

In May 1957, the United States and the Soviet Union exchanged proposals for a test ban and cutoffs in the production of materials for nuclear weapons but failed to reach an agreement. In March 1958, the Soviet Union announced a unilateral suspension of testing after completing its latest series of many atmospheric nuclear explosions. It urged the United States to do likewise. It came just as the United States was about to start a major series of weapons tests in the Pacific and at the Nevada Test Site. On April 8, 1958, President Eisenhower proposed a technical conference to explore the verification of a test ban.

Many people are not aware that Eisenhower and his administration were very involved in several proposals and negotiations to ban or limit nuclear testing. President Kennedy is often cited for achieving, as he did in 1963, the Treaty Banning Nuclear Weapons Tests in the Atmosphere, in Outer Space, and Under Water, often called the Limited Test Ban Treaty (LTBT). It did not cover underground testing. Nevertheless, much work on testing limitations and verification occurred during Eisenhower's presidency.

EARLY ATTEMPTS TO IDENTIFY NUCLEAR TESTS

The first Soviet test in 1949 resulted in an expansion of a variety of monitoring technologies by the United States, Great Britain, and Canada. These included seismic stations, infrasound detectors for low-frequency acoustic waves in the atmosphere, hydrophones for detecting explosions in the oceans, and instruments for detecting explosions in the upper atmosphere and space.

In July and August 1958, a Conference of Experts to Study the Methods of Detecting Violations of a Possible Agreement on the Suspension of Nuclear Tests was held in Geneva. It was often called the Conference of Experts. Western delegations were from the United States, the United Kingdom, Canada, and France; eastern delegates were from the USSR, Czechoslovakia, Romania, and Poland. James Fisk, a senior scientist and administrator at Bell Labs, headed the U.S. delegation; William Penny, the British. Supporting the U.S. delegation were a number of advisers, including five seismologists and a number of scientists familiar with nuclear weapons programs. Yevgeni Fedorov, a Soviet geophysicist, headed the eastern delegation. Unlike the U.S. delegation, the USSR delegation included a senior diplomat. Many criticized Eisenhower for not including a senior diplomat in the U.S. delegation.

Their agenda was unclear. In his 2009 book *Detecting the Bomb: The Role of Seismology in the Cold War*, U.S. seismologist Carl Romney wrote, "Fedorov promptly raised an essentially political issue by proposing a statement of objectives that amounted to a prior commitment to a test ban." Romney also stated, "Fisk countered that the Conference should confine its work to defining methods of detection, analyzing their capabilities and limitations, and let Governments decide how to use the information."

In advocating a treaty first with verification later, the Soviet Union stated that monitoring posts on its territory staffed by foreign citizens would be an excuse for espionage. The United States wanted strong verification measures such as monitoring stations in the USSR. The Soviet delegation received new instructions on many of the verification proposals enumerated by Fisk. Informal meetings on each of these topics were largely technical and constructive, whereas formal meetings of the entire delegations were more politically charged.

A final report completed in August 1958 contained agreed-upon conclusions on verifiability, the establishment of a global network of 180 monitoring (control) stations, and the need for on-site inspections. Nevertheless, identifying underground explosions and distinguishing their signals from those of earthquakes were the most contentious issues, as they remained for many decades. I think that these issues, in fact, were resolved by 1970.

Much data existed by 1958 on monitoring explosions in the atmosphere but not much on underground tests. Detection of acoustic waves from explosions in the oceans was known to be considerably easier than identification of underground nuclear explosions. In 1955 the United States detonated a 30-kiloton underwater device called *Wigwam* in the eastern Pacific. Sensors detected this and the *Baker* underwater test of 1946 at large distances. The Soviet Union conducted two underwater nuclear explosions of 6 and 10 kilotons near its Arctic islands of Novaya Zemlya. Romney stated that both underwater shots were well recorded by stations of the U.S. Atomic Energy Detection System (AEDS). Those results, the locations of stations, and the capabilities of AEDS at the time, however, were classified and remained so for decades. While Romney had access to them, most other scientists did not.

The experts from the United States and the USSR agreed that earthquakes under the oceans, which account for about half of the world's earthquakes, could be considered as identified if they did not produce large acoustic signals (sound waves) in the oceans. An explosion in the water column is much more efficient at generating those waves than an earthquake within the oceanic crust.

Many earthquakes deeper than about 30 to 45 miles (50 to 75 km) could be identified using the seismic wave pP, which arrives soon after the P wave and is reflected from the surface of the Earth near the earthquake source (figure 3.1). These included deeper earthquakes beneath the Kuril-Kamchatka region, the most active area of the USSR, and those beneath the Hindu Kush and Pamir mountains of Central Asia. Shallower earthquakes are often difficult to identify using pP waves because they arrive very soon after P waves.

When the Conference of Experts met in August 1958, all nuclear tests had been conducted either in the atmosphere or underwater with the exception of a single very small U.S. underground explosion code-named *Rainier*. It was detonated in a volcanic rock called tuff during September 1957 at the Nevada Test Site, which had opened in 1951. Its yield of 1.7 kilotons was about ten times smaller than the nuclear explosions of

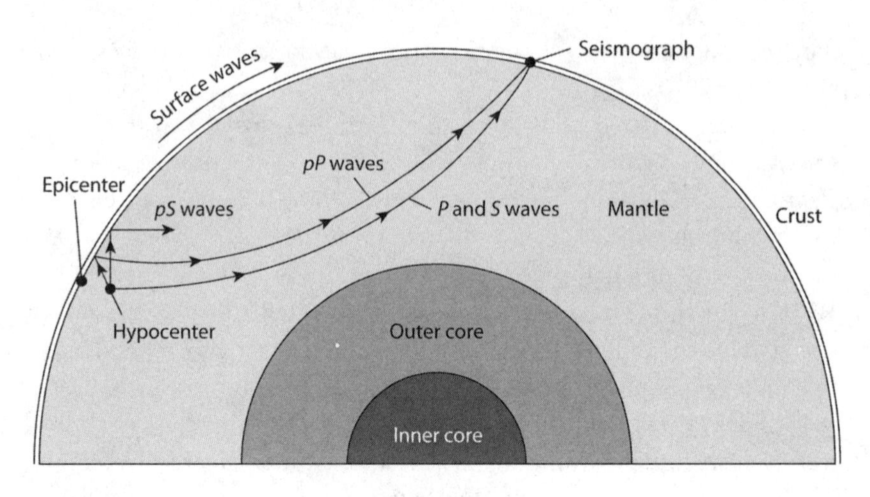

FIGURE 3.1

Interior of the Earth. Lines with arrows indicate paths of seismic waves between a source, such as an earthquake, and a recording station. The epicenter is the point on Earth's surface directly above the hypocenter of an event.

Source: Sykes and Evernden, 1982.

1945 and ten thousand times smaller than the largest hydrogen bombs tested as of 1958.

U.S. government scientists twice reduced their estimate of *Rainier*'s seismic magnitude because stations at different distances gave different results. Many people contested the idea that the 1957 test was difficult to detect seismically and that its signals were hard to distinguish from those of earthquakes. The USSR did not detonate an underground test until October 1961. It was of similar size to *Rainier* and was likely detonated at the Eastern Kazakhstan test site in much harder rock than *Rainier*. It was well detected outside the USSR.

Throughout 1958, both the Soviets and Americans set off many nuclear explosions. The Soviet Union conducted nuclear tests in the atmosphere from February until November 3, 1958. Several were probably in the megaton range. The United States conducted a series of nuclear explosions in the Pacific, called Operation Hardtack I, starting in April 1958. It consisted of atmospheric explosions with yields ranging from a few kilotons to about 10 megatons (10,000 kt) and two underwater shots.

Testing ended in August 1958 with two high-altitude nuclear explosions with yields of 3.8 megatons near Johnston Island in the Pacific, which produced radio blackouts and observable effects in the Pacific similar to northern lights.

The United States conducted additional nuclear explosions in the atmosphere at the Nevada Test Site in September and October 1958 as part of Operation Hardtack II, which also included underground tests of *Logan* at 5 kilotons, *Blanca* with a yield of 22 kilotons, and several explosions smaller than one kiloton.

Upon completion of the Experts' Report at the end of their conference on August 22, 1958, President Eisenhower proposed a one-year testing moratorium on all testing if the Soviet Union also refrained. This moratorium would be extended on a year-by-year basis if the control system recommended by the Experts could be installed and progress made on arms control agreements. While a treaty was being negotiated, both the Soviet Union and the United States continued to conduct nuclear tests into the fall of 1958.

Treaty negotiations, called the Conference on the Discontinuance of Nuclear Tests, began in Geneva on October 31, 1958, a day after the *Blanca* explosion by the United States. Parties to the conference consisted of the three nuclear states at the time—the United States, the United Kingdom, and the Soviet Union.

The USSR's second underground nuclear test did not occur until February 1962. It took place at its Eastern Kazakhstan test site with a yield comparable to that of the U.S. test *Blanca*. Because most of Russia consists of areas of old geologic units that favor very good propagation of seismic waves, the United States overestimated the yields of underground nuclear explosions at the two main Soviet test sites, including those of 1961 and 1962, by a factor of about three times and continued to do so for decades.

The United States calibrated unknown Soviet yields from their measured seismic magnitudes using measured yields and magnitudes of underground nuclear explosions in Nevada and southern Algeria. This turned out to be a big mistake that led to the large overestimations of Soviet yields. The Soviet Union did know, however, that seismic waves propagated efficiently in various parts of their country from its earlier large chemical explosions.

U.S. REASSESSMENT OF DIFFICULTIES WITH VERIFICATION IN 1959

A key factor in the U.S. reluctance to negotiate a treaty was that U.S. Department of Defense officials and others were not convinced about the reliability of identifying nuclear explosions of moderate and small size. They argued that seismic signals from underground explosions based on 1958 data were both smaller and harder to distinguish from the signals of earthquakes of comparable size than the forecasts of the Experts in August 1958.

After reviews within the U.S. government, these new findings from underground explosions in 1958 were tabled as negotiations resumed in Geneva in January 1959. The White House issued a public statement about the new findings and their implications. Because some of the best seismic data and information for the underground explosions in Nevada in 1958 were classified, it was difficult, however, for most seismologists to form independent judgments about the identification of small nuclear explosions.

The relevance of the new U.S. data on underground explosions was strongly challenged by Soviet geophysicists Y. Riznichenko and Leonid Brekhovskikh in an article in *Pravda* on January 20, 1959. The Soviet ambassador to the conference stated that the Experts' report of the previous summer should be the sole technical basis for the negotiations. The U.S. ambassador to the Geneva talks insisted that the new data must be considered. He also proposed a technical meeting to discuss detection of high-altitude nuclear explosions.

The Eisenhower administration formed a Panel on Seismic Improvement, consisting of U.S. seismologists and other experts, that was chaired by Lloyd Berkner, president of Associated Universities. Known as the Berkner Panel, its initial assignment was to ascertain if improvements could be made to the monitoring stations proposed by the Experts so that the number of problem seismic events per year being considered would not have to be increased. Their main task was to resolve the difficulties in identifying underground explosions and distinguishing their seismic signals from those of earthquakes.

The findings of the Berkner Panel were introduced at the Geneva conference in June 1959. Their recommended improvements consisted of greater numbers of seismic sensors at each control post, the use of seismic surface waves for discriminating earthquakes from explosions, and the placing of unmanned instruments, so-called black boxes, in earthquake-prone areas of the USSR and the United States. The black boxes would be designed to be tamper proof.

SEISMIC MAGNITUDES USED TO DESCRIBE THE SIZES OF EARTHQUAKES AND UNDERGROUND NUCLEAR EXPLOSIONS

It is necessary to have a simple way to describe the sizes of earthquakes. Seismologists use many different magnitude scales to do this. They all involve measuring the amplitude, or size, of a particular seismic wave, taking the logarithm of its amplitude, and making a correction for the distance between the source and a seismic station. Logarithms are used because the sizes of seismic waves vary over a huge range. Because magnitude scales are logarithmic, magnitude increases one unit when amplitude increases ten times. Hence, an earthquake of magnitude 6 is ten times larger in amplitude than an event of magnitude 5. In 1935, Charles Richter at Cal Tech devised the original magnitude scale, called M_L or Richter magnitude, for local earthquakes in southern California. The magnitudes listed in table 3.2 are not those proposed by Richter but ones appropriate for events at large distances.

TABLE 3.2 **Earthquake Magnitudes**

m_b	Determined from the amplitudes of short-period seismic P waves, usually at distances greater than about 1250 miles (2000 km)
Ms	Determined from the amplitudes of long-period seismic surface waves
m_{bLg}	Determined from the amplitudes of seismic waves at regional distances (less than 1250 miles or 2000 km) within continents
Mw	Determined from very long-period (low-frequency) seismic waves; called moment magnitude

In describing underground nuclear explosions and earthquakes, the United States advocated the use of the seismic magnitude m_b, which could be measured from P-waves on seismograms, whereas explosive energy or yield could not be measured directly. Energy is a physical quantity, whereas seismic magnitude is not.

FURTHER PROPOSALS BY THE EISENHOWER ADMINISTRATION

In the 1959 negotiations, the United States also emphasized that methods could be used to mask or reduce the sizes of seismic waves from nuclear tests. Disagreements with the Soviet Union about evasion and the determination of seismic magnitudes were a complete nonstarter for the remainder of the talks with the USSR through 1963.

In April 1959, Eisenhower proposed a phased approach to achieving a comprehensive nuclear test ban. He stated that Soviet proposals did not provide for effective control (i.e., verification) and recommended starting with a ban on nuclear tests in the atmosphere. The proposal was endorsed by British prime minister Macmillan and, with some further modifications, was accepted by Soviet premier Nikita Khrushchev. This made it likely that a test ban treaty could be signed at the Paris summit that both President Eisenhower and Premier Khrushchev agreed to attend in May 1960.

The downing of an American U-2 spy plane over the Soviet Union on May 1, however, led to an atmosphere of hostility that cut short the Paris summit and the chance for a test ban during the remainder of Eisenhower's presidency. Glen Seaborg, the discoverer of plutonium for which he won the Noble Prize, a chairman of the Atomic Energy Commission, and an adviser to ten U.S. presidents, was known for daily entries in his diary. In his 1981 book he wrote, "President Eisenhower, alarmed by the increased tensions, was reported to have decided to order the resumption of testing if Richard Nixon won the election" [in November 1960]. After Kennedy's victory, he advised the president-elect "to resume testing without delay."

The Geneva conference recessed in December 1959 with each of the three parties submitting annexes about the verification of underground testing. Because the Soviet annex differed considerably from those of the United States and Britain, a considerable impasse existed. Negotiations resumed in January 1960 with proposals for annual quotas on the number of on-site inspections and a threshold ban on underground testing, which was to prohibit large underground tests. The United States proposed a seismic magnitude threshold of 4.75, about 15 to 20 kilotons. This conversion of magnitude to yield was based solely on a few underground tests in 1957 and 1958 in Nevada in tuff, a soft rock. A magnitude m_b of 4.75 corresponds to smaller nuclear explosions at the two main Soviet test sites.

The USSR declared a moratorium on nuclear testing after completing a large series of nuclear tests in November 1958. Likewise, the United States followed its own self-declared moratorium after completing the *Hardtack* tests in October 1958. Neither country is known to have tested nuclear explosions for nearly three years. One exception is that the United States conducted explosions in secret with tiny nuclear yields equivalent to the explosive power of only several pounds (kilograms) of TNT. Called *hydronuclear*, these explosions confirmed that U.S. weapons were "one-point safe"—that is, a sudden shock at one point on their surface would not lead to a large inadvertent nuclear explosion.

In the meantime, France conducted its first nuclear test in Algeria in February 1960. The Soviet Union stated that the United States could obtain information about the development of nuclear weapons from the French even if the United States was observing a moratorium. This provided an excuse when the USSR resumed nuclear testing in September 1961.

SOVIET SERIES OF LARGE NUCLEAR TESTS IN 1961 AND 1962

The Soviet Union suddenly ended its self-declared moratorium on testing, detonating a huge series of nuclear explosions starting on September 1, 1961. The previous day, it announced that it would resume testing. It came at the height of the Berlin crisis, when the Soviet Union attempted to block western access to that city and the United States flew in supplies. Several tests were in the megaton range, and about six were 10 to 30 megatons. Numerous Soviet tests, most of them in the atmosphere, continued until December 25, 1962.

The Russians also detonated the largest nuclear explosion ever conducted at their Arctic test site near the islands of Novaya Zemlya on October 30, 1961. Premier Khrushchev announced it ahead of time. The explosion occurred in the atmosphere with a yield of about 42 to 50 megatons, more than three thousand times the power of the Hiroshima bomb of 1945 (figure 3.2). It is often called either "the Tsar bomb" or "Big Ivan."

Its detonation certainly scared the United States government. Soviet nuclear scientists later said it was never weaponized; it was one of a kind and was not mass-produced for delivery by airplanes or missiles. In fact, the development of new nuclear weapons in the mid-1960s to the mid-1970s involved yields smaller than several megatons for single-warhead missiles and yields smaller than a megaton for missiles with multiple warheads.

The Soviet resumption of nuclear testing caught the United States by surprise. The many tests must have been planned well in advance. Some people in the United States accused the Soviet Union of cheating by

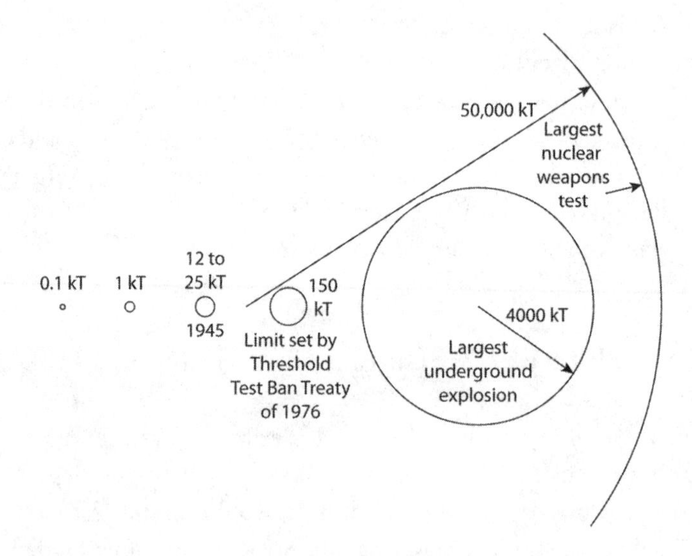

FIGURE 3.2

Yields of key nuclear explosions detonated by the Soviet Union and the United States in the atmosphere and underground. Size or yield of nuclear explosions is in kilotons (kt).

Unpublished figure by the author.

preparing to test while negotiations were in progress in Geneva. Both sides, however, were following self-declared moratoria at the time, not treaties, protocols, or other agreements of indefinite duration. Nevertheless, it contributed to a U.S. view that the Russians could not be trusted.

President Kennedy ordered the resumption of nuclear testing by the United States on September 5, 1961; explosions commenced in Nevada ten days later. The United States and the United Kingdom started testing at Christmas Island in the Pacific in April 1962 and then at Johnston Island.

In 1957 the Soviet Union used a large missile to launch a satellite called *Sputnik*. Its success was followed by the failure of a U.S. rocket launched from the ground that was to deploy a satellite. This led to apprehension in the United States that the USSR was ahead in the development of intercontinental ballistic missiles (ICBMs) that could carry nuclear weapons.

President Eisenhower, however, was well aware of the secret development of military ICBMs by the United States, but he did not publicly divulge its existence. Kennedy ran for election as president in 1960 on the claim that a "missile gap" favored the Soviet Union. The apparent gap disappeared soon after Kennedy became president. The United States observed via early satellite imagery that the Soviet Union possessed fewer ICBMs than was claimed. The United States was, in fact, ahead in deployed ICBMs.

THE CUBAN MISSILE CRISIS

The world came the closest it has come thus far to nuclear war during the Cuban Missile Crisis of October 1962. An American U-2 spy plane secretly photographed nuclear missile sites being built in Cuba by the Soviet Union. Decades later senior Soviet military officials revealed that their commanders in Cuba had been authorized to launch medium-range SS-4 ballistic missiles at the United States if the United States attacked the Russian sites in Cuba. The SS-4's carried nuclear weapons. Khrushchev undertook the installation of missiles in Cuba, apparently regarding Kennedy as young and inexperienced.

Several high-ranking military officials advised Kennedy to invade Cuba, but fortunately he refrained. Instead, he instituted a U.S. naval blockade to prevent the Soviets from bringing more military equipment and supplies into Cuba. Both Kennedy and Khrushchev recognized the devastating possibility of a nuclear war and publicly agreed to a deal. The Soviets withdrew their nuclear weapons and missiles from Cuba. Under an unannounced part of the agreement, which remained secret for more than twenty-five years, the United States withdrew old Jupiter missiles from Turkey.

A few months after the crisis, a personal insight came from my father, who was in charge of security at the Washington Air Traffic Control Center. He was responsible for keeping military aircraft separated from civilian flights in their broad area of traffic control. He mentioned to me the huge numbers of military flights over the southeastern United States during the Cuban crisis and said, "You don't know how close we came to war."

THE LIMITED NUCLEAR TEST BAN TREATY OF 1963

The Cuban Missile Crisis and the fear of nuclear war became a turning point in test ban negotiations. Khrushchev wrote Kennedy in December 1962, "The time has come now to put an end once and for all time to nuclear tests." He offered three on-site inspections (OSIs) per year. The United States informed Soviet officials in February 1963 that it was willing to reduce its proposed annual quota of OSIs to seven. Many people assumed that a compromise between three and seven OSIs would be easy to achieve and a test ban would follow. Nevertheless, different views in the United States about evasive testing, the effectiveness of monitoring, and the use of seismic magnitudes continued to impede an agreement that would include underground tests.

A conciliatory speech by Kennedy at American University in June 1963 and Khrushchev's welcoming of it led to an agreement among the United States, the United Kingdom, and the USSR to hold test ban talks in Moscow. Former ambassador to the Soviet Union W. Averell Harriman led the U.S. delegation. A full CTBT was the U.S. objective. If that was not possible, Harriman was to seek a limited treaty banning nuclear tests in the atmosphere, underwater, and in space. On the opening day of the negotiations in July 1963, the USSR submitted a draft treaty that included those environments but not underground testing. The Soviet Union would accept unmanned "black boxes" on its territory but not on-site inspections. OSIs were considered essential by the United States.

Discussion of a full test ban did not extend beyond the first day of the negotiations. The three principals very quickly agreed to a limited test ban treaty (LTBT), which was signed on August 5, 1963. It prohibited any explosion that "causes radioactive debris to be present outside the territorial limits of the State under whose jurisdiction or control such explosion is conducted." The treaty was opened to signature by all states three days later and entered into force on October 10, 1963.

The treaty was of indefinite duration but with an escape clause for a national security emergency. Each country could conduct monitoring using its own means, so-called national technical means (NTM), including satellite surveillance. The treaty included neither on-site inspections nor seismic stations on the territories of the others.

SUPPORT FOR AND OPPOSITION TO THE LIMITED TEST BAN TREATY

As urgent as it was to pass agreements curbing the nuclear arms race, serious dissent was expressed in both the Eisenhower and Kennedy administrations about not only a Comprehensive Nuclear Test Ban Treaty (CTBT) but also a Limited Test Ban Treaty (LTBT). In June 1963, President Kennedy held a vigorous debate at a meeting of his Committee of Principals for Nuclear Testing about the positions the United States should take in upcoming test ban negotiations with the Soviet Union and Britain. That debate occurred in the wake of the Cuban Missile Crisis of late 1962.

Much of the material and several quotes that follow are from Glen Seaborg's 1981 book. (I felt honored in the 1980s when he congratulated me at a meeting of the U.S. National Academy of Sciences for my work on nuclear verification.) According to Seaborg, Secretary of Defense Robert McNamara "supported the draft treaty because he felt that the United States was ahead and that a test ban would freeze our superiority." Jerome Wiesner, Kennedy's science adviser, said "the [weapons] laboratory directors were not in a position to judge policy considerations."

Secretary of State Dean Rusk said he thought the draft treaty, which included underground testing, "was consistent with all of the interests of the country, including national security." Rusk went on to state "it was his impression that the Principals had all agreed that the risks to national security from an unlimited arms race were greater than the risks from a test ban treaty." According to Seaborg, "Rusk doubted that any amount of discussion would ever bring the technical people to agree among themselves."

On behalf of the Joint Chiefs of Staff, General Maxwell Taylor stated that under the western draft treaties of August 1962, the Soviet Union could make important gains in weapons development through clandestine testing. Taylor also said the Joint Chiefs would likely prefer a limited ban with a continuation of underground testing. According to Seaborg, the Joint Chiefs relied heavily on the directors of the weapons laboratories for their information. A number of dissenters were from those labs. Seaborg said that McNamara referred to statements made by R. W. Henderson of the Sandia Laboratory and John Foster, director of the Livermore weapons lab, to the effect that U.S. warheads could not penetrate to Soviet targets unless further tests were undertaken to correct defects. Foster continues to oppose a full test ban today.

Seaborg reported that at the end of June 1963 Kennedy was pessimistic about a comprehensive test ban treaty (CTBT), based mainly on the impasse about the attributes and number of on-site inspections. One senator reported to Kennedy in 1963 that a CTBT likely would be ten votes shy of the mandatory sixty-seven (two-thirds) needed for U.S. Senate approval of the treaty. According to Seaborg, the exclusion of underground tests from the Moscow talks in 1963 was all but sealed when Khrushchev accused the West on July 2 of demanding on-site inspections for espionage purposes.

Seaborg regarded "the failure to achieve a comprehensive test ban as a world tragedy of the first magnitude." He went on to state, "While I did not take this position at the time [1963], looking back I tend to agree with those who feel that our concern about Soviet cheating was exaggerated . . . it is doubtful that clandestine tests the USSR might have undertaken in violation of a comprehensive treaty would have been militarily significant in the aggregate."

Seaborg stated that Kennedy threw himself into the ratification process. In contrast, President Clinton's involvement was largely lacking when the Senate defeated the Comprehensive Test Ban Treaty in 1999 after allowing little time for testimony or debate. In anticipation of the Senate debate and vote in 1963, Kennedy was adroit in having a bipartisan group of senators attend the signing ceremony for the Limited Test Ban Treaty (LTBT) in Moscow. Kennedy did not want to see the treaty defeated by Republicans and southern Democrats nor have it go down to defeat like the Versailles Treaty under President Woodrow Wilson.

During the Senate debates on the limited test ban in 1963, Secretaries Rusk and McNamara testified in favor of it. Seaborg pledged that the Atomic Energy Commission "would continue under the treaty to support vigorous research and development programs in its weapons laboratories and would thus be able to retain able scientists and engineers, attract new ones, and maintain the vitality of the laboratories." General Taylor recommended four safeguards that the Joint Chiefs of Staff thought were necessary: (1) a continuing program of underground testing, (2) maintenance of modern nuclear laboratories, (3) the ability to resume atmospheric testing if required, and (4) improved monitoring capabilities. He concluded that if those safeguards were established, the risks inherent in the treaty could be accepted through a stabilization of international relations.

Edward Teller, often called the father of the H-bomb, gave several reasons why he opposed the LTBT: (1) the treaty involved a field that had repeatedly proved itself unpredictable; (2) it would prevent the United States from acquiring information about weapons effects needed to design defenses against incoming ballistic missiles; (3) it would stimulate, not subdue, the arms race; (4) it would not deter proliferation; and (5) it would seriously wound the U.S. Plowshare program for peaceful uses of nuclear explosions. He argued that the Soviets had acquired knowledge of weapons they would need for defense against incoming ballistic missiles during their nuclear tests of 1962.

Several prominent scientists and administrators disagreed with Teller on weapons for defense against incoming ballistic missiles. For example, Norris Bradbury, the director of Los Alamos, characterized the treaty as "the first sign of hope that international nuclear understanding is possible." But John Foster, director of Livermore, had serious reservations about the LTBT and reiterated a number of the points made by Teller. Lewis Strauss, former chairman of the Atomic Energy Commission under Eisenhower, saw no advantage to the treaty.

General Thomas Power, commander of the Strategic Air Command, also expressed strong opposition to the Limited Test Ban Treaty. He said the security of the United States depended on having overwhelming superiority over the Soviet Union and that atmospheric testing was needed to achieve that superiority. He and some others seemed to yearn for a prior era in which superiority and massive retaliation were possible U.S. policies. Soviet diplomats stated that they would not allow the United States to have the superiority it possessed in 1962 during the Cuban Missile Crisis. Although the USSR had not reached parity with the United States in nuclear weapons and their delivery systems in 1963, the passage of the LTBT could not prevent the Soviet Union from inflicting unprecedented and immense amounts of damage and loss of life on the United States in a major exchange of nuclear arms.

Prior to the full floor debate by the Senate, President Kennedy wrote to both the majority and minority members of the Senate pledging to implement the four safeguards recommended by the Joint Chiefs of Staff. Two key senators, Everett Dirksen and Henry Jackson, who were expected to vote against the treaty, came out in favor of it. On September 24, 1963, the Senate approved the Limited Test Ban Treaty by a vote of 80 to 19.

POSITIVE AND NEGATIVE CONSEQUENCES OF LTBT

The Limited Test Ban Treaty of 1963 had both positive and negative effects. It led to a great reduction in radioactive fallout and pollution from atmospheric tests. In that sense, it was a successful public health measure and an environmental accomplishment. Because nuclear fallout was eliminated, interest in a full test ban nearly ceased. Nevertheless, many people hoped the LTBT would act as a brake on the arms race and would lead to further arms control agreements, including a full test ban, and to nuclear disarmament.

The LTBT did not put an end to the detonation underground of progressively larger numbers and yields of nuclear weapons. The arms race continued with the development of new weapons, tested underground rather than in the atmosphere. By the late 1960s, the USSR and the United States had conducted underground tests of hundreds of kilotons. Without the LTBT, however, it is possible that each would have conducted even larger tests in the atmosphere.

The LTBT was the first treaty with the Soviets that involved nuclear arms. The Cuban Missile Crisis led the superpowers to negotiate a treaty that they thought would be a step toward preventing nuclear war. Other treaties followed, albeit with long lag times, involving reductions in intercontinental and intermediate-range (300 to 3500 miles, or 500 to 5500 km) missiles, weapons to knock down ballistic missiles, and nuclear weapons in outer space.

U.S. nuclear doctrine changed from "massive retaliation" in Eisenhower's presidency to what was termed "flexible response" during the Kennedy administration. The LTBT came at a time of transition from great U.S. superiority in nuclear weapons and intercontinental delivery systems to approximate parity with the Soviet Union. Emphasis after 1963 focused on specific arms control agreements and not on general disarmament.

Seaborg stated, "It was always the view of Kennedy and his advisors that a comprehensive test ban would be far more effective than a limited test ban in preventing the proliferation of nuclear weapons." Neither France nor China, which developed and tested nuclear weapons for the first time in the early to mid-1960s, signed the LTBT. They continued testing in the atmosphere until 1974 and 1980, respectively. Both then tested solely underground until just before they signed the CTBT in 1996.

A CTBT in 1963 probably would not have deterred France and China from continuing to develop and test nuclear weapons. India signed the LTBT and did not conduct a nuclear test until 1974, which it claimed was for peaceful purposes. India might have been persuaded not to acquire nuclear weapons if a full test ban had been enacted in 1963. Its greatest concern in 1974 was not so much Pakistan as the nuclear capabilities and military strength of China. If India had not tested, it is possible that Pakistan would not have developed and tested nuclear weapons. Neither India nor Pakistan, nor later North Korea, tested in the atmosphere, only underground. The LTBT may have encouraged the Treaty for the Prohibition of Nuclear Weapons in Latin America, which entered into force in

1968. Since then, treaties have established nuclear-free zones in several other regions.

A Comprehensive Test Ban Treaty in 1963 would have halted the development of weapons for new long-range missiles, but this was not to be. At that time, the United States possessed only single-warhead missiles such as Minuteman I and Polaris. Likewise, in 1963 the Soviet Union only had intercontinental missiles with single warheads. The United States, followed by the Soviet Union, developed multiple independently targetable reentry vehicles (MIRVs) for long-range missiles. A MIRVed missile carries more than one reentry vehicle and its nuclear warhead. Each of those warheads can be sent independently to separate targets, hence the name MIRV.

The sum of the nuclear yields on a MIRVed missile is about half the yield on a single-warhead missile of similar size. Each MIRVed missile, however, is more dangerous in that its several warheads can be detonated over a larger area, causing more destruction than a large warhead carried by one missile. A single large warhead on a non-MIRVed missile, though twice as large, expends much of its energy higher in the atmosphere, not near the surface of the Earth. With the development of MIRV and more accurate missiles, somewhat smaller yield weapons largely replaced megaton-size warheads. The yields of individual warheads on land-based and submarine-based MIRVed missiles of the United States and the USSR are in the range 50 to 750 kilotons.

In the early 1970s, the Soviet Union tested new warheads for its MIRVed missiles. Clearly, a full test ban in 1963 or somewhat later would have prevented that testing and deployment. The same is true for the MIRVed warheads on the U.S. Minuteman III, Peacekeeper (MX), Poseidon C3, Trident I C4, and Trident II D5 missiles.

MIRVed missiles were one of the most dangerous aspects of the nuclear arms race between the United States and the USSR, especially as missile accuracy improved. They have not been eliminated and are still very dangerous today. Several warheads on one MIRVed missile could be aimed independently at several missiles of the other superpower. Because MIRVed nuclear missiles could be used in a first strike, they put each of the two superpowers in the very dangerous posture of launching its missiles very quickly in response to a warning that may have been a false alarm.

A full test ban in 1963 could have prevented the development and testing underground of nuclear weapons for missile defense, as in the very large U.S. Amchitka explosions *Milrow* in 1969 and *Cannikin* in 1971 and advanced Soviet anti-ballistic weapons. Weapons for cruise missiles and MIRVed intermediate-range missiles, such as the Soviet SS-20, would not have been developed and tested.

The *Cannikin* warhead was designed several years before it was detonated in 1971. It was to be part of the Safeguard anti-ballistic missile (ABM) system, which was abandoned almost two years before the *Cannikin* test for reasons of cost and technical difficulties. Much public outcry arose understandably about stationing ABM systems with very large nuclear warheads near U.S. cities. The British newspaper the *Guardian* stated just before the *Cannikin* test, "When President Nixon canvassed the opinion of seven Government agencies, his own Office of Science and Technology pointed out that the test was of only marginal technical usefulness. Of the seven agencies canvassed only two, the U.S. Atomic Energy Commission and the Department of Defense, expressed approval." The costs of carrying out the *Milrow* and *Cannikin* tests at remote Amchitka Island in the western Aleutians must have exceeded a billion dollars.

4

ATTEMPTS TO HIDE NUCLEAR TESTS: THE BIG-HOLE EVASION SCHEME

One of the major issues related to monitoring a full nuclear test ban involves the possibility of evasive testing, particularly what is referred to as decoupling or muffling the seismic signals of an underground nuclear explosion. Sometimes also called the big-hole hypothesis, its vast overexaggeration was critical in excluding underground tests from the Limited Test Ban Treaty in 1963.

Most nuclear explosions were conducted without any attempts to muffle or decouple their seismic waves. They were fully coupled or tamped events in which the surrounding rock was in close or nearby contact with the nuclear device. They produced extremely large permanent (nonelastic) rock deformation, including vaporization of rock near the shot point. In contrast, full decoupling involves keeping the rock surrounding a large cavity in the elastic domain so that no permanent deformation occurs.

In 1959 Edward Teller, a controversial physicist who was then at the Lawrence Livermore National Laboratory and had worked on the U.S. nuclear weapons program during and after World War II, and Albert Latter of the Rand Corporation, a nonprofit institution that advises the executive branch of the U.S. government, argued that seismic signals from underground nuclear explosions could be greatly reduced—that is, greatly muffled—by detonating them in large underground cavities. They convinced many people in the United States during the late 1950s and 1960s that monitoring could not keep up with the ability to either evade detection or disguise the seismic signals from nuclear explosions.

This turned out not to be the case; verification clearly kept ahead of evasion for decades. Misstatements about decoupling, whether driven by those who were poorly informed or others who purposefully misled, were made again in the 1999 Senate debate about ratification of the Comprehensive Nuclear Test Ban Treaty (CTBT). Three senators argued in that debate that decoupling could allow huge nuclear explosions to be hidden from U.S. and international monitoring. The decoupling concept was not new; it had been around since 1959, contrary to what those senators implied in 1999. As I explain later in this chapter, a gigantic cavity at a considerable depth in the Earth is needed to decouple or muffle even a small nuclear explosion, plus there is the need to insure that its radioactive products do not leak to the surface of the Earth where they could be detected. These are formidable obstacles for a potential evader.

Latter's 1959 estimates of the amount of muffling of seismic waves achieved by a decoupled underground explosion were based on data from the single very small U.S. underground explosion in 1957 called *Rainier*. It actually was not muffled; unlike the hard rock at the two main Soviet test sites, from which seismic waves propagate efficiently, *Rainier* was detonated in relatively soft rock in Nevada and produced small seismic signals.

Far too many policy makers in the United States in the 1960s placed too much credence on the work as well as on the arguments and testimony of Latter and Teller. Both men were major proponents of the concept that successful decoupling was possible and strong advocates of continued nuclear testing. Latter and his associates concluded that seismic waves from *Rainier* would have been smaller by a factor of forty to fifty if it had been conducted in a large underground cavity at depth in the same rock. They predicted an amplitude reduction of three hundred times relative to *Rainier* for a fully decoupled underground explosion in salt deposits, which turned out to be quite incorrect.

In joint congressional testimony in 1960, Representative Chester Holifield asked, "Let us understand what that means. Does that mean that a 300-kiloton shot could be reduced in seismic recordings to a 1 kiloton recording?" Latter answered, "Yes, sir." Senator Gore went on to ask, "Do you agree with that, Dr. Romney?" Seismologist Carl Romney replied, "Yes, indeed."

Carl Romney was the lead seismologist working for the Air Force Technical Center (AFTAC) of the Department of Defense. AFTAC operates the U.S. classified (secret) monitoring system and the data analysis center for detecting and identifying nuclear tests by other countries. (In later chapters I discuss my involvement and controversy on several occasions with Romney about monitoring nuclear explosions and determining yields of Soviet explosions.) Although it is not very evident in his 2009 book *Detecting the Bomb: The Role of Seismology in the Cold War*, Romney was very conservative and incorrect in his assessments of monitoring the Soviet Union for many decades. He and Teller were extremely distrustful of arms control agreements with the Soviet Union and of the effectiveness of nuclear monitoring.

Many U.S. public officials in the 1960s were left with the misguided impression that it was possible to construct a cavity suitable to fully decouple an explosion of 300 kilotons and that such a nuclear explosion would not be detectable even with much improved seismic networks. Unfortunately, the issue continues to be contentious today.

Within only about a year of his proposal in 1959, Latter and a few others presented the decoupling concept as a well-developed and well-tested hypothesis at a joint U.S.-USSR-UK conference in Geneva and in hearings by the Joint Committee on Atomic Energy of the U.S. Congress. Nevertheless, publication of the decoupling theory and data from the *Cowboy* chemical experiments in salt to test it did not occur until 1961.

Most federal agencies did not possess the scientific or technical expertise to evaluate the decoupling hypothesis, even with appropriate security clearances and the need to know. Most of those with purported expertise on decoupling were in the weapons labs, mainly Livermore.

The calculations and inferences about decoupling by Latter and his associates from 1959 through the 1960s were incorrect and quite misleading for a number of reasons. They overestimated the amount of muffling by a large factor. They made extrapolations of explosions to different depths in the Earth and to other rock types without any basis in actual nuclear testing. They also used simplified assumptions about the containment of radioactive products. They assumed incorrectly that rock materials were uniform and without cracks, joints, and other imperfections that were familiar to geologists and geophysicists. Latter stated that muffling factors as large as two to three thousand could also be achieved by lining a large underground cavity with an absorbing material like carbon, even though U.S. tests incorporating that concept with two small nuclear explosions proved otherwise.

Based on no experimental observations, Latter stated in 1959 that the propagation of seismic waves from underground explosions in salt should be poorer than those from *Rainier*, which was conducted in tuff, a light, porous rock formed by the consolidation of volcanic ash. When the United States resumed testing in late 1961, it conducted its first peaceful nuclear explosion, *Gnome*, of 3 kilotons underground in a thick salt deposit in southeastern New Mexico.

Both the United States and the Soviet Union conducted peaceful explosions to form cavities in salt for storage of various products, breaking tight rocks for petroleum extraction and other purposes described later. *Gnome* was fully coupled, not muffled, and its seismic waves, in fact, were large, not small compared to *Rainier*'s as Latter had wrongly deduced. Seismic amplitudes were especially large for paths to stations to the east of *Gnome* in the United States. Seismic waves recorded for the 5.3-kiloton *Salmon* explosion set off later in salt in Mississippi on October 22, 1964, also were large at seismic stations in eastern and central North America.

These areas acted as a good comparison for the more easily detected efficient propagation of seismic waves in similar old and strong rocks of the crust and uppermost mantle of the Earth beneath most of Russia. Similarly, the seismic amplitudes and magnitudes of many later USSR nuclear explosions in salt were large and easily detected compared to explosions of similar yields in softer rock in Nevada.

President Eisenhower chose James Killian, the president of MIT, as his science adviser in late 1957 and looked to him for scientific guidance about arms control proposals. Killian was aided by scientists appointed to the new President's Science Advisory Committee (PSAC), many of whom stressed the value of a test ban in preventing fallout from atmospheric explosions as a first step in controlling the arms race.

This helped to counter the negative views about nuclear arms control of Admiral Lewis Strauss of the Atomic Energy Commission; Edward Teller; Admiral Arthur Radford, chairman of the Joint Chiefs of Staff; and many others in the Department of Defense. As an example of some of the rhetoric during this era, Radford stated in May 1957, "We cannot trust the Russians on this [nuclear testing] or anything." Those who still take that view today, sixty years later, are unlikely to change their minds about a Comprehensive Nuclear Test Ban Treaty even in the face of huge improvements in verification.

In his 1977 book *Sputnik, Scientists, and Eisenhower*, Killian, a pragmatist who would have been classified politically as a moderate or Rockefeller Republican, discussed the "big hole" or decoupling concept of evasion in his role as Eisenhower's science adviser. He states, "Teller wished to make a dramatic demonstration of the possibilities of cheating, and this [decoupling] was it." Killian goes on to note that "the Berkner panel heard Latter's theories about the big hole, and in its report concluded that decoupling techniques existed which could reduce the seismic signal by a factor of ten or more. . . . The big-hole technique proved to be much more difficult than expected by its advocates. . . . It was a bizarre concept, contrived as part of a campaign to oppose any test ban."

Killian continued, "I was asked by the State Department to lead an American technical delegation to London to give the British the information about the 'big hole' and other methods of concealing nuclear tests. . . . While we were in London, Dr. Latter said to me in casual conversation that whatever advances might be made in detection technology, the West Coast group led by Teller would find a technical way to circumvent or discredit them."

In discussing the process of scientific review and advice to the government, Killian said, "As Henry R. Myers has written, this is true even today. There seems to be a widely held obsession with the *possibility* of violations rather than with their probability, or their significance. . . . Opponents of limitations on nuclear testing have exploited this obsession by encouraging fears that have little basis in fact. We should have strengthened the campaign for a test ban by making clear when an apparent technical question is not really technical. . . . We who spoke for science never succeeded in making clear the difference between probability and possibility."

The hypothesis of decoupling was confirmed in general during the *Cowboy* experiments of 1959, in which chemical explosions of up to one ton (0.9 metric tons) were set off in small cavities excavated in salt in Louisiana. Decoupling or muffling factors of one hundred were calculated for salt soon afterward, using the *Cowboy* data and theoretical calculations. Nearly twenty years after the *Cowboy* experiments, however, it became known that the decoupling effectiveness for those chemical shots had been overestimated. The amounts of decoupling for *Cowboy* were revised downward by 30 percent to account for new information on the energy release associated with the detonation of the unconfined high explosive, Pelletol, which was used in the cavity explosions for *Cowboy*, as compared to the energy release of the same explosives when they were confined.

This revision to a factor of seventy brought the muffling factor from the *Cowboy* experiments into agreement with the decoupling factor of seventy obtained for the U.S. *Sterling* underground nuclear explosion of 0.38 kilotons (380 tons) in 1966. *Sterling* was detonated in the cavity in Mississippi produced by the 1964 *Salmon* nuclear explosion of 5.3 kilotons in salt.

Since 1966, several people working on evasive testing have continued to extrapolate the *Cowboy* and *Sterling* data to yields as large as 10 to 100 kilotons. Clearly, given the political implications for monitoring a CTBT, much better peer review of the original *Cowboy* data and a similar experiment by independent groups should have been made. Peer review also would have been in order for the early estimates of the amounts of muffling claimed by Latter and others.

I decided to work on decoupling around 1985, because it was so critical to the success of a future CTBT. I discussed muffled nuclear testing in a long paper I wrote in 1996, "Dealing with Decoupled Nuclear Explosions Under a Comprehensive Test Ban Treaty." John Murphy of Science Applications International Corporation (SAIC), a proponent of the belief that other countries were able to conduct large decoupled explosions evasively, described my 1996 paper as "the other view." Hans Bethe of Cornell University, one of the most important physicists of the twentieth century and an arms control advocate, however, was very complimentary about my paper, which greatly pleased me.

Actual data on decoupled (muffled) nuclear explosions are very meager because little testing was undertaken and most information on it is thirty to forty-five years old. The 2012 National Academies Report states that in the era of nuclear testing prior to early 1996, nuclear experiments by the United States to test this scenario were not considered important enough to be given priority and financial support except at very small yields.

The present nuclear decoupling database includes the following:

1. Only one fully decoupled nuclear explosion, 0.38-kiloton *Sterling*, detonated by the United States in Mississippi in 1966
2. A partially decoupled explosion, *Azgir 3-2*, by the Soviet Union at Azgir in western Kazakhstan in 1976, 8 to 10 kilotons
3. *Mill Yard* in 1985, about 0.02 kilotons (20 tons), and three other very small nuclear explosions at the Nevada Test Site (NTS)

The *Sterling* and *Azgir 3-2* nuclear explosions were conducted in cavities in large salt domes created some time earlier by much larger, fully coupled nuclear explosions—*Salmon* of 5.3 kilotons and *Azgir 3-1* of 64 kilotons. Salt domes are large bulbous structures of nearly homogeneous salt formed by the instability of less dense salt rising upward through more dense sedimentary rocks.

A huge cavity in salt is needed to fully decouple a nuclear explosion of 5 kilotons (figure 4.1). The depth range for containment of fully decoupled explosions of various yields in salt is illustrated in figure 4.2. At shallow depths, nuclear explosions are not contained and either blow out to the surface, form large craters, or leak detectable radioactive materials (upper blue area). The minimum depth for containment increases with yield.

Salt, however, is unlike most common rocks in that it deforms readily at low confining pressures and shallow depths in the Earth's crust. A stable air-filled cavity in salt becomes unstable and deforms severely or collapses at depths greater than about 3000 feet (900 m). Thus, the depth range in the Earth for conducting a fully decoupled nuclear explosion in salt is quite limited. The 1966 *Sterling* and the 1976 *Azgir 3-2* explosions were detonated at nearly optimum depth, about 3000 feet, to insure containment and cavity stability. If those explosions had been conducted at shallower depths, larger cavities of greater volume would have been needed to obtain the same amount of decoupling.

The USSR's 1976 *Azgir* test is very important for estimating the detection and identification capabilities of decoupled nuclear explosions larger than one kiloton because its yield was about twenty-three times larger than that of the U.S. *Sterling*. This 1976 explosion, however, was about three times larger than that required for full decoupling to occur. Hence, it is not surprising that it was only partially decoupled, with

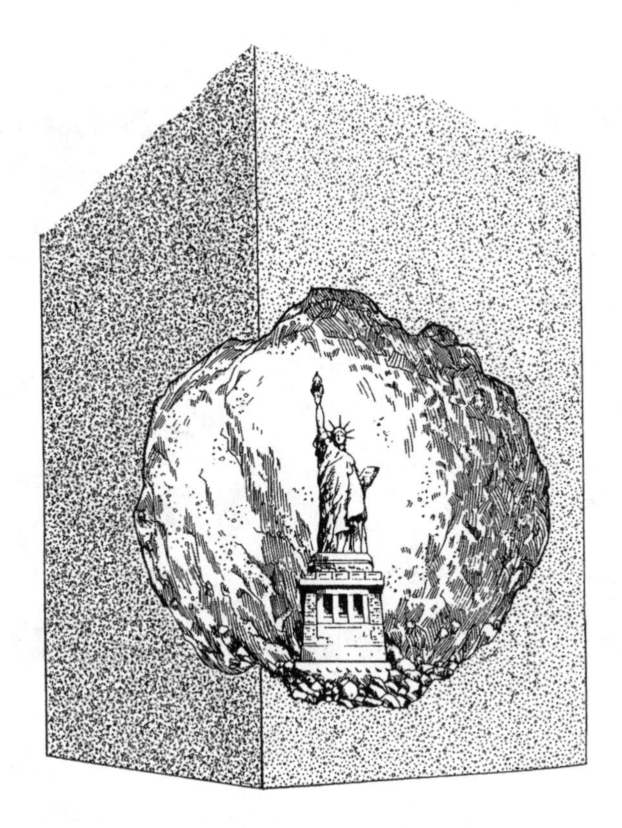

FIGURE 4.1

To fully decouple a 5-kiloton nuclear explosion in salt requires a spherical cavity of diameter 282 feet (86 m) at a depth of about 3000 feet (900 m). It would be larger than the Statue of Liberty and its pedestal.

Source: Office of Technology Assessment, 1988.

seismic waves muffled by twelve to fifteen times, as I reported in my 1996 paper, not the seventy times associated with full decoupling. The cavity in salt in which the 1976 *Azgir* explosion was detonated was huge, with a mean diameter of 243 feet (74 m). It was large enough to contain the Statue of Liberty and its pedestal. Its seismic magnitude m_b of 4.06 was well recorded by stations as far away as Canada.

Nevertheless, on May 10, 1976, *Newsweek* reported that the 1976 Soviet *Azgir* explosion had been promptly identified as a fully decoupled nuclear test. The yields mentioned in that article on it and the 1971 nuclear test

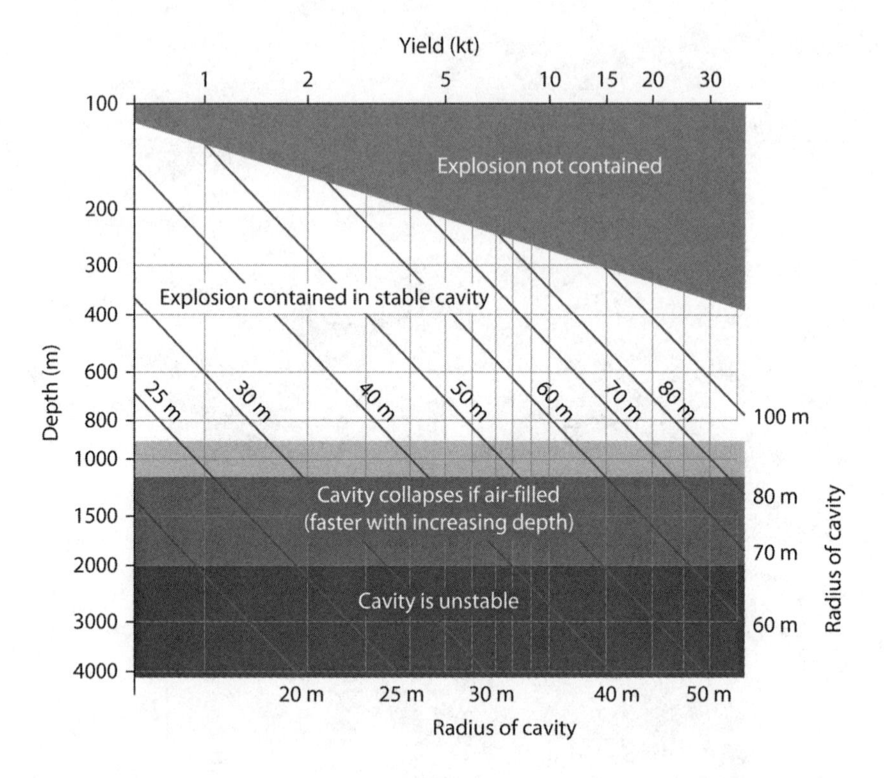

FIGURE 4.2

Depths (vertical axis) for conducting decoupled nuclear explosions in an air-filled cavity in salt. Yield of the explosion is on the horizontal axis at top; the cavity radius is at bottom.

Modified from Davis and Sykes, 1999.

that created its cavity, however, were not corrected for seismic magnitude bias (described earlier). *Newsweek*'s yields were much too large—210 kilotons for 1971 and 40 to 50 kilotons for the 1976 test, not 64 kilotons as published by the USSR for 1971 and 8 to 10 kilotons that I calculated for 1976 by including my estimate of seismic bias.

As I stated in my 1996 paper, "The *Newsweek* article conveys the impression that the USSR was capable of conducting a decoupled test of 40 to 50 kt, not as we now know, only a partially decoupled test of 8 to 10 kt." News media don't just make up a story with details like that.

Some person or agency must have provided that information. Again, the U.S. overestimation of Soviet yields had major implications for national security policy, in this case a large overestimate of what the Soviet Union could achieve by evasive nuclear tests.

Decoupled testing requires an air-filled or an evacuated cavity. Very large cavities have been constructed in salt by conventional and solution mining, mainly for storage of oil, gas, compressed air, and waste products. Solution mining is the least expensive method of forming a large cavity at depth in salt. It involves using a pipe to pump fresh water from the surface deep into salt. A mixture of salt and water called brine is then pumped back out through another pipe.

Disposal of brine is a major problem, requiring about seven times the volume of water to extract one volume of brine using solution mining. A brine-filled cavity is not suitable for conducting a decoupled explosion because an explosion in brine or water is well coupled, producing large seismic waves. Hence, brine must be pumped out of a cavity and disposed of in secret if it is to be used for decoupled testing.

Solution-mined cavities are also very fragile. Very large brine-filled cavities rarely have been pumped out, leaving them empty. The U.S. Strategic Petroleum Reserve is stored in large cavities in salt created by solution mining at Bryan Mound, Louisiana. Brine in those cavities is typically replaced by oil. When oil is removed later, the cavity is filled back up with brine. Part of their support comes from the brine or oil in them, each of which, of course, is much denser than air. Emptying them and not replacing oil by brine or water is strictly discouraged because these cavities are expensive, fragile structures that may collapse when emptied. The technical literature describes a number of large cavities in salt that have fully or partially collapsed.

Relatively little is known about the strength of salt surrounding a cavity produced by solution mining. The *Sterling* decoupled and the *Azgir* partially decoupled explosions were conducted in cavities in salt formed by previous, much larger nuclear explosions. Much of the energy in a fully decoupled nuclear explosion goes into creating very high pressure in the cavity, about two hundred times atmospheric pressure at a depth of about 3000 feet (900 m). Less energy goes into forming seismic waves, which is why such an explosion is, in fact, decoupled. The walls of the cavities created by the two nuclear explosions were subjected to strong heating. Hence, the properties of the salt at and near those cavity walls are likely to be very different from the salt surrounding a cavity produced by solution mining.

A weaker solution-mined cavity may well collapse or deform significantly if a decoupled test is conducted in it, considerations that do not seem to have been taken into account by L. A. Glenn of Livermore, William Leith of the U.S. Geological Survey, Larry Turnbull of the CIA, and others who have long maintained that an air-filled, solution-mined cavity in salt could be used for decoupled nuclear testing with yields up to 10 kilotons.

The Pre-Caspian depression, which contains the world's largest concentration of salt domes, including those at Azgir, is now located mostly in the independent country of Kazakhstan. That salt extends into the Russian Republic near the mouth of the Volga River. Bedded salt—salt deposits that have not been deformed naturally into salt domes—is found to the north of Lake Baikal in the Russian Republic. Bedded salt typically is not as favorable as salt domes for constructing large cavities because the salt layers usually are interbedded with other sedimentary rocks.

Another large area of salt domes is found near the northern coast of the Gulf of Mexico in the United States. Widespread salt deposits are also found in China and Iran, limited deposits in Pakistan, very limited quantities in India, and no known salt deposits in North Korea, a country of very old crustal rocks. Large volumes of water are needed to produce a solution-mined cavity in salt, but many parts of the Middle East and North Africa from Iran through Libya receive very little rainfall per year. Obtaining enough water for significant solution mining in those areas is highly problematic. Iran contains many large salt structures, but about half the country consists of deserts that receive less than 10 inches (25 cm) of rain per year. Higher rainfall is found in the Zagros Mountains of western Iran, but they are well monitored by seismic stations in adjacent countries.

U.S. NUCLEAR EXPLOSIONS IN VERY SMALL CAVITIES

Mill Yard, set off in 1986, and three other very small U.S. nuclear explosions were detonated in hemispherical cavities in soft rock at the Nevada Test Site with radii of about 36 feet (11 m). One publication estimated

Mill Yard's yield as about 20 tons (0.02 kt). It may have been decoupled but by an unknown amount. Purging the air in the tunnel used for *Mill Yard* released 5.9 curies of radioactive materials into the atmosphere, which would be easily detected today.

Tiny Tot, another test of very small yield in a hemispherical cavity, was detonated in granite at the Nevada Test Site (NTS) in 1965 to provide information on the effects on hard rock of a surface nuclear burst. The amount it was decoupled has not been published and may not be known. Data from such an early very small underground test are likely to have been sparse. In 2013 Australian physicist Christopher Wright reported that, starting fifteen minutes after the explosion, an uncontrolled release of predominantly noble gases, particularly xenon, emanated from the mouth of the *Tiny Tot* shaft.

These very small explosions in hemispherical cavities apparently were not intended as tests of full decoupling, but they may well have been partially decoupled. These examples from Nevada indicate that the containment of bomb-produced radioactive products, especially noble gases, is problematic.

CHEMICAL EXPLOSIONS IN CAVITIES IN SALT AND HARD ROCK

Given the scarcity of data for decoupled nuclear explosions, a number of individuals and groups have attempted since 1959 to estimate decoupling factors using chemical explosions in cavities in salt and hard rock. Their explosive yields, however, ranged from less than a ton to about ten tons, much smaller than the 380-ton yield of the small U.S. *Sterling* decoupled nuclear explosion.

Several of those experiments were poor for one or more of the following reasons: (1) chambers containing the explosive were not sealed and were open to the outside; (2) the explosives used were old munitions; (3) tamped (fully coupled) explosions were not included in the experiments as comparisons; and (4) the explosives were placed on the floor, not in the center of a cavity. Suspending several tons of explosive near the center of a cavity is not easy. The U.S. *Cowboy* chemical explosions in salt cavities in 1959 and Soviet chemical explosions in cavities in limestone in Kirghizia in 1960, with explosive yields up to six tons, are some of the better experiments. Salt is the only rock in which a decoupling factor as large as seventy has been obtained for chemical explosions in underground cavities.

Hard rocks are much more common than salt. Constructing a large cavity at depth in hard rock, however, is much more difficult and expensive than one in salt. In 1995 F. A. Heuzé of the Livermore Lab and others examined hard rocks in terms of their suitability for decoupled testing. They found that a salient characteristic of hard rocks is that they are seldom massive, monolithic formations, but rather are penetrated by numerous cracks, faults, and other discontinuities, which may provide pathways for radioactive leakage from decoupled tests. To my knowledge, no one has either measured released xenon and other gases or determined if they can be contained for a decoupled nuclear explosion in hard rock. In addition, it is difficult to characterize how hard rock will respond to the strong shock and high pressure of a decoupled nuclear explosion based on traditional methods of laboratory tests on small rock samples.

Heuzé and others stated that the igneous rock in which the *Piledriver*, *Tiny Tot*, and *Hardhat* nuclear explosions were conducted at the Nevada Test Site was not a granite of good quality. Joints in it are spaced about 8 inches (20 cm) apart. Joints spaced about 3 feet (1 m) characterize the hard rocks of the French test site in the Hoggar massif of southern Algeria. Hence, one or more joints could well leak radioactive products following a decoupled test in hard rock.

In 1992 I attended a session on the containment of bomb-produced radioactive materials at an open meeting on decoupling. Most of the participants, many of whom were experts on containment, thought that a large cavity in hard rock would need to be extensively reinforced to prevent collapse when a nuclear explosion was detonated in it.

CLAIMS OF EVASIVE DECOUPLED TESTING BY THE SOVIET UNION

In 1995 Larry Turnbull of the CIA wrote that nuclear explosions had been conducted evasively by the Soviet Union in mines, one in 1972 on the Kola Peninsula and a second in the Ukraine on September 16, 1979. The first claim is clearly contradicted by published information, and the second is likely false.

In his 1975 review of Soviet peaceful nuclear explosions (PNEs), Milo Nordyke of the Livermore Lab described a proposed ore-breaking project using a 1.8-kiloton PNE. A Soviet list contains a 2.1-kiloton explosion on September 4, 1972, on the Kola Peninsula in a well-known mining area. Forty-seven open stations recorded it with a magnitude of 4.6. All indications are that it was well coupled, not muffled. Soviet geophysicists did measure seismic amplitudes on either side of a slit cut into the rock, which may have led Turnbull to claim incorrectly that it was a decoupled test.

In 1992 the *New York Times* reported a nuclear explosion of 1/3 kiloton at noon on September 16, 1979, in a Ukrainian mine. Sultanov and others list it as occurring in sandstone within a coal mine with a yield of 0.3 kiloton. Using that location, Frode Ringdal of Norway and Paul Richards of Lamont computed an origin time at noon Moscow time and a magnitude of 3.3 using signals received at the Norsar seismic array near Oslo. It would be even better recorded and located today. Its somewhat smaller magnitude for its yield is reasonably attributed to the explosion's being conducted in soft rock, not to decoupling in hard rock.

U.S. MEETINGS ON DECOUPLING IN 1996 AND 2001

In 1996 I was invited to attend a classified meeting on clandestine nuclear testing organized by the U.S. Arms Control and Disarmament Agency (ACDA) and the Defense Special Weapons Agency (DSWA). DSWA was formerly the Defense Nuclear Agency and later became the Defense Threat Reduction Agency. ACDA was subsequently merged with the State Department.

Because ACDA did not have enough funding for the 1996 meeting, DSWA provided funding, but in exchange it largely controlled the agenda. The meeting was mainly a forum for presentation of work by one consulting group, Jaycor, which worked under a contract from the predecessor to DSWA, the Defense Nuclear Agency. Jaycor included a number of former employees of DNA.

Speakers from Jaycor claimed that important nuclear testing could be carried out evasively by Iran, Libya, and North Korea. The meeting, however, did not provide an opportunity to evaluate fairly the prospects for detecting evasive testing. Most of the rest of us not connected with Jaycor were only able to ask questions or make short remarks. Jaycor proposed a site in Iran, which they called "El Cheato," and claimed that Iran could use it for clandestine nuclear testing. My sense was that Jaycor's knowledge of evasive testing was poor. For some reason they did not use AFTAC's classified information for that site, even though they had access to it. Calling a site in Iran El Cheato might have been considered cute if many people in the audience had not taken them seriously.

In 2001 I participated in a forum at the secret level on decoupled testing organized by the State Department's Bureau of Verification, Livermore Lab, and the Department of Energy. Several people who had worked on decoupling for many years, such as Lew Glenn of the Livermore Lab, claimed that large decreases in seismic wave amplitudes could be produced by large decoupled tests. Glenn, unrestrained by the moderator, interrupted me and other speakers repeatedly. William Leith of the U.S. Geological Survey reiterated his previous statements about the possibility of using very large holes in the ground for decoupled nuclear testing. With only three slides allotted to me, I chose to speak about the *Azgir* partially decoupled explosion of 1976.

The State Department itself had no expertise on decoupled testing—neither its physical basis nor geological and containment constraints—until it hired Robert Nelson a decade later. Little was accomplished at that meeting to narrow the issues involved with evasive testing.

I contend that decoupling is no longer a problem today for nuclear explosions of military significance. Huge cavities would be needed to fully decouple nuclear explosions with yields of one to a few kilotons. I return later to claims about decoupled nuclear testing during the Senate hearings in 1999 on the Comprehensive Test Ban Treaty and the 2012 Report by the National Academies.

5

U.S. OVERESTIMATION OF SIZES OF SOVIET UNDERGROUND EXPLOSIONS: 1961–1974

One of the key issues that greatly intensified the concerns of many of us in the scientific community was how the United States determined the yield—the size or energy release—of Soviet underground nuclear explosions. For many decades, the U.S. government calculated Soviet yields based on seismic data and yields from explosions detonated in Nevada, a procedure that led to a major overestimation of the yields of Soviet explosions.

A factor in the yield question in the late 1950s and 1960s concerned how many earthquakes occurred in the USSR whose seismic magnitude equaled that of a given yield (size) of an underground test. These numbers were important at the time of the negotiations for a full test ban treaty in 1963 because the seismic signals of many small earthquakes could not be distinguished from those of small underground nuclear explosions.

INCORRECT DETERMINATION OF YIELDS OF SOVIET UNDERGROUND EXPLOSIONS

Two factors affected the determination of the yields of Soviet underground nuclear explosions:

1. Seismic magnitudes vary with the types of rocks in which explosions of a given yield are detonated.
2. Seismic magnitudes also vary because seismic P waves travel easily or poorly beneath various testing areas.

As a result of these two factors, several values of yield can be calculated for the magnitude of a single underground explosion.

I make use here of the seismic magnitude m_b (Table 3.2) to calibrate the yield or energy release of Soviet underground nuclear explosions. Later I will use the surface wave magnitude Ms as well.

VARIATION OF YIELD WITH ROCK TYPE

Prior to the resumption of nuclear testing by the USSR in October 1961, the United States proposed prohibiting underground tests of seismic magnitude m_b of 4.75 and larger. This magnitude is determined from the first-arriving, short-period P waves from either an explosion or an earthquake. At the time, the United States took an m_b of 4.75 to be equivalent to a yield of about 15 to 20 kilotons. It was widely thought that the seismic signals from smaller explosions in the Soviet Union could not be distinguished from those of small earthquakes.

Originally, however, U.S. estimations of Soviet yields from magnitudes were based solely on underground tests conducted in 1957 and 1958 in Nevada in tuff, a soft rock formed by consolidation of volcanic ash, and on data from a test site characterized by poor propagation of P-waves to large distances. Very small magnitudes were computed for those early U.S. underground nuclear explosions for their known yields, which led to biased and incorrect determinations of Soviet yields. It is understandable that the Soviet Union, with hard rocks at its test sites, was not interested in a ban on underground tests based on a magnitude threshold.

Donald Springer of Livermore and colleagues indicated in 2002 that the United States tested nuclear explosions underground in a great variety of rock types at the Nevada Test Site (NTS) but overwhelmingly in two soft rocks, tuff and alluvium. The United States conducted very few tests at NTS in hard rocks like granite and only a few in rocks of intermediate strength like dolomite and rhyolite.

Alluvium consists of unconsolidated sand and gravel. Unlike hard rocks, it often can be dug with a shovel and has very low seismic wave speeds. It is not a suitable material in which to construct a large cavity for decoupled (muffled) testing because a cavity in it would collapse quickly. U.S. nuclear explosions in alluvium were set off in two environments—dry and water saturated. Tests in dry alluvium generated the smallest seismic magnitudes of any rock type for a given yield. This is understandable because much energy is expended in the closure of air spaces between sand and gravel.

Very few places in the world—the Nevada Test Site (NTS) and Namibia, but none in Russia—have thicknesses of dry alluvium suitable for conducting tests larger than one or two kilotons. Figure 5.1 shows many collapsed craters in alluvium in Yucca Valley, the site of many U.S. tests within NTS. Those craters were (and still are) easily visible on satellite imagery.

FIGURE 5.1

Aerial view of craters produced by underground nuclear explosions in the Yucca Flat portion of the Nevada Test Site. View is looking from south to north.

Source: Springer and colleagues, 2002.

The United States conducted two tests of about one kiloton, called *Unde* and *Ess*, in alluvium in 1951 and 1955. Their depth of burial, 16 and 62 feet (5 and 19 m), was so shallow, however, that they produced subsidence craters at the surface. The much-discussed *Rainer* test in 1957 was detonated at a greater depth of 900 feet (274 m) in a tunnel in tuff. *Blanca* in 1958, with a yield of 22 kilotons, was detonated at 990 feet (301 m) in very soft tuff that had a low seismic wave speed and low strength. Springer and others indicated that the cavity it produced collapsed twelve seconds later. Tuff clearly is also not appropriate for the construction of a large cavity at depth in the Earth.

When the United States resumed testing in late 1961, it detonated the nuclear explosion *Fisher* of 13.4 kilotons in dry alluvium at the Yucca testing area at a depth of 1194 feet (364 m). It too produced a collapsed cavity at the surface. In 2009 Carl Romney of the Department of Defense reported that *Fisher*, when adjusted for differences in yield, was forty times *smaller* in seismic amplitude than the underground explosion *Gnome* in salt in New Mexico. Clearly, for the same yield, large differences in the seismic magnitude m_b were associated with explosions conducted in different rock types.

In 1974 seismologist Romney of the Defense Department urged the United States government to base the maximum size of underground tests under the Threshold Test Ban Treaty (TTBT) on seismic magnitude. Yield, not magnitude, however, was accepted as the threshold in 1974 by the two parties to the treaty—the United States and the Soviet Union. Soviet negotiators had argued that yield, or energy release, was a physical quantity, whereas magnitude was not. That was correct. If magnitude had been the threshold, it would have allowed the United States to conduct explosions of much larger yield at the Nevada Test Site in tuff or alluvium than the Soviet Union could do at their two test sites in old, much harder rock.

VARIATION OF SEISMIC MAGNITUDE WITH PROPAGATION OF SEISMIC P WAVES IN THE UPPER MANTLE OF THE EARTH

From 1958 through 1988, the U.S. government did not take into account a second important factor that affects the measured seismic magnitude, m_b, of underground explosions of the same yield at the Nevada Test Site and the two main Soviet test sites. Seismic waves were not absorbed as much beneath those Soviet test sites at depths of 30 to 125 miles (50 to 200 km) as they were beneath the Nevada Test Site. The amount of absorption is related to temperatures at those depths. NTS experienced much younger volcanism and its associated heating than the geologically older rocks beneath the two main Soviet test sites. Soviet scientists had presented data on chemical explosions of several kilotons that indicated larger magnitudes than those the United States had measured for explosions of the same size in Nevada. Still the U.S. government ignored those results.

For more than a decade, Romney and Eugene Herrin of Southern Methodist University, both seismologists, advocated the use of an incorrect method to calculate yields of Soviet underground explosions. Based on known yields and seismic magnitudes, m_b, of explosions at NTS, the magnitudes of Soviet explosions were used to calculate unknown Soviet yields. U.S. estimates of Soviet yields were too large not only because of differences in rock type but also because the propagation of P waves beneath Soviet test sites was very efficient. These differences are illustrated in figure 5.2. P waves crossing parts of the upper mantle of the Earth, called the asthenosphere, where temperatures are high, become reduced in amplitude.

During the Reagan administration, accusations by the U.S. government that the Soviet Union was cheating on the Threshold Test Ban Treaty were based on these incorrect calibrations and indirectly on the belief that the Russians could not be trusted. The U.S. yield estimation procedures were classified "secret" for decades. Hence, those lacking the appropriate clearances could not judge those yield calculations, making it difficult to prove that Romney's and Herrin's methods were erroneous. British and many U.S. scientists, including me, who worked on test ban verification, did not agree with the U.S. procedures for determining Soviet yields. Incorrect calculations also led to claims in the United States that the yields of deployed Soviet weapons were larger than they actually were by a factor of about three times.

A variety of evidence indicated that the Earth's uppermost mantle beneath Nevada differed from regions of older geology such as those found beneath the two main Soviet test sites and beneath the central and

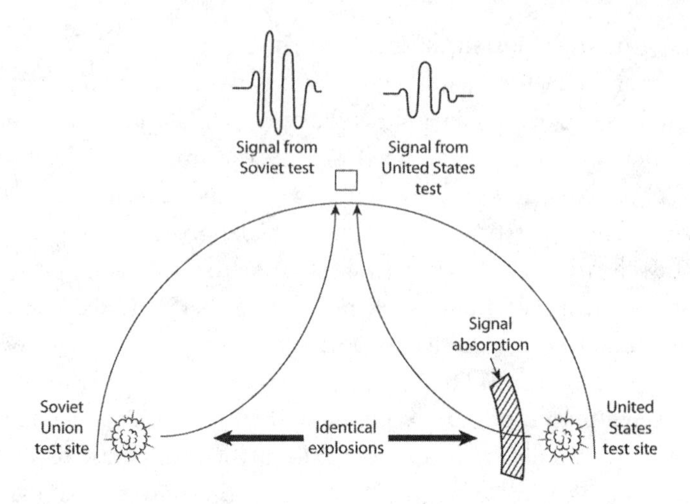

FIGURE 5.2

Illustration of differences in sizes (amplitudes) of seismic P waves for underground nuclear explosions of the same yield at Soviet and U.S. test sites. Seismic P waves from explosions at the Soviet Union's two main test sites are much larger than P waves from an explosion of the same yield in Nevada.

Source: Office of Technology Assessment, 1988.

eastern United States. I described earlier this effect for the *Gnome* and *Salmon* explosions in New Mexico and Mississippi. In 1967 Jack Evernden, then at the Air Force Technical Application Center (AFTAC), used many observations of seismic waves to show that P-wave propagation and P-wave speeds were very different beneath Nevada and the central and eastern parts of the United States. Evernden's work and the observations from *Gnome* and *Salmon* should have been a warning not to use data from Nevada to calibrate yields of explosions at Soviet test sites.

In his 2009 book, Romney accepted major differences between the propagation of seismic waves in the western United States and regions of older crust and uppermost mantle. Nevertheless, I know he was an ardent foe of this view for many decades, especially at many classified meetings. The control of classified seismological data by a few persons, such as Romney and Herrin, prevented a resolution of this issue between 1974 and 1988. Because I was very involved in those deliberations, at both the classified and open levels, I knew their views well throughout that long debate.

Romney also stated that, unknown to seismologists outside AFTAC, the Soviets fired their first underground nuclear explosion on October 11, 1961. He went on to say that six open seismic stations reported signals to the U.S. Coast and Geodetic Survey (USCGS) for their unclassified estimate of location and depth for the 1961 event. Romney says the USCGS location was near the Eastern Kazakhstan test site (which was correct) but their depth estimate of 19 miles (31 km) was not. USCGS, a government agency, was not permitted to state whether the event was an explosion or an earthquake, which it did not.

Most seismologists, who lacked access to classified AFTAC data, were not able to voice an opinion about the nature of the 1961 event. When seismic stations are located at very large distances, as must have been the case for those open stations used by USCGS, uncertainties in estimating depth can be large, i.e. 30 miles (50 km). Without access to the AFTAC data, identifying the event based solely on open data as either an earthquake or a nuclear explosion would have been poor seismological practice. Hence, the depth reported by the USCGS was so uncertain that the event could not be identified from open data alone.

In a chapter published in the book *Soviet Nuclear Weapons*, by Cochran and others, in 1989, Steven Ruggi and I estimated the yield of the small 1961 Soviet test as about one to a few kilotons. We knew of its existence because it was included in a table published in a book by Bruce Bolt of UC Berkeley in 1976. The second Soviet underground test, in February 1962, whose yield Ruggi and I estimated in 1989 as 10 to 20 kilotons, was well recorded by many stations outside the USSR. Our yield was similar to that of the U.S. underground nuclear explosion *Blanca* of 1958.

Thus, the U.S. classified AEDS program was able to detect and locate a Soviet nuclear explosion as small as one to a few kilotons as early as 1961, but with estimates of yields that were higher than those by Ruggi and me. An accurate and less biased U.S. method for calibrating yields of those early Soviet underground explosions might have had a major impact on the test ban negotiations in 1962 and 1963. Nevertheless, the Soviet Union was not helpful to the process because it released neither the yields of those two explosions nor seismic data from stations within their country.

6

NEW METHODS TO IDENTIFY UNDERGROUND TESTS: 1963–1973

Although large seismic events could be identified in 1963, few methods were available then to distinguish, or discriminate, the signals of small underground nuclear explosions from those of the many small earthquakes that occur worldwide every day. More specifically, this discrimination was needed for countries of special interest to the United States at the time—the Soviet Union and China.

In the 1950s, the U.S. government quickly recognized seismology's potential for detecting and identifying underground nuclear tests. A panel of technical experts headed by physicist Lloyd Berkner recommended in 1959 that the United States should greatly expand funding of seismology to increase fundamental understanding and to develop better instrumentation.

Subsequent funding for the underground explosion part of the program came from what was called the Vela Uniform program. Run by the Advanced Research Project Agency (ARPA) of the Defense Department, it transformed seismology almost instantaneously from a sleepy, poorly supported scientific backwater to a field flooded with new funds, instruments, professionals, students, and excitement. Seismology and its instruments for recording earthquakes became the main technology for detecting, locating, and identifying underground nuclear tests. As a result of this investment, seismologists today can identify much smaller seismic events than they could in 1963.

DETERMINATION OF DEPTHS OF EARTHQUAKES AND EXPLOSIONS

No nuclear explosions have been set off deeper than 3 miles (5 km), and nearly all have been detonated within the upper 2 miles (3 km) of the

Earth's crust. Therefore, the reliable determination that a seismic event is considerably deeper means that it almost certainly cannot be a nuclear explosion. Holes drilled deeper than 6 miles (10 km) in hard rock are difficult to use for clandestine testing. An experimental deep Soviet scientific test hole on their Kola Peninsula deformed quickly in response to large stress differences.

Good determinations of depths greater than about 30 miles (50 km) readily identified events beneath the Kuril Islands (figure 6.1) and Kamchatka, the most active areas of the USSR, as earthquakes. Similar determinations also identified frequent deep events beneath the Hindu Kush and Pamir mountains of Central Asia as earthquakes.

Many earthquakes with depths shallower than 30 miles (50 km) occur well offshore of the Kamchatka Peninsula and the Kuril Islands of Russia (figure 6.1). They can be distinguished as earthquakes by their very small hydroacoustic (sound) waves in the Pacific Ocean. Explosions within the water column generate much larger acoustic waves that propagate readily to large distances in the oceans. Hence, earthquakes offshore of, and most earthquakes beneath, Kamchatka and the Kuril Islands have been identifiable as such for many decades.

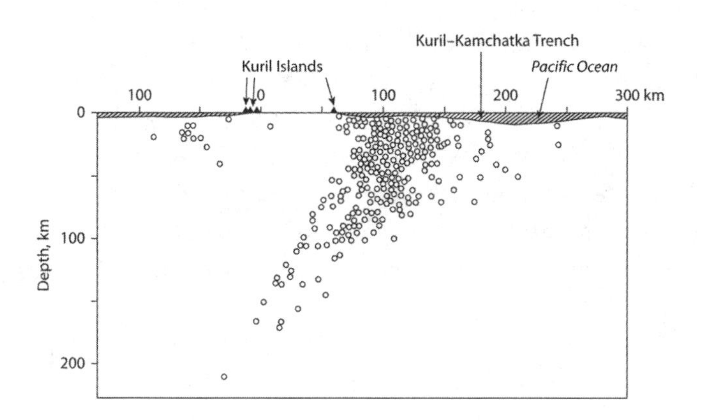

FIGURE 6.1

Locations of earthquakes projected onto a cross section of the Earth extending from the Pacific Ocean at the right, across the Kuril subduction zone, to the eastern Sea of Okhotsk at the left.

Source: Sykes, Evernden, and Cifuentes, 1983.

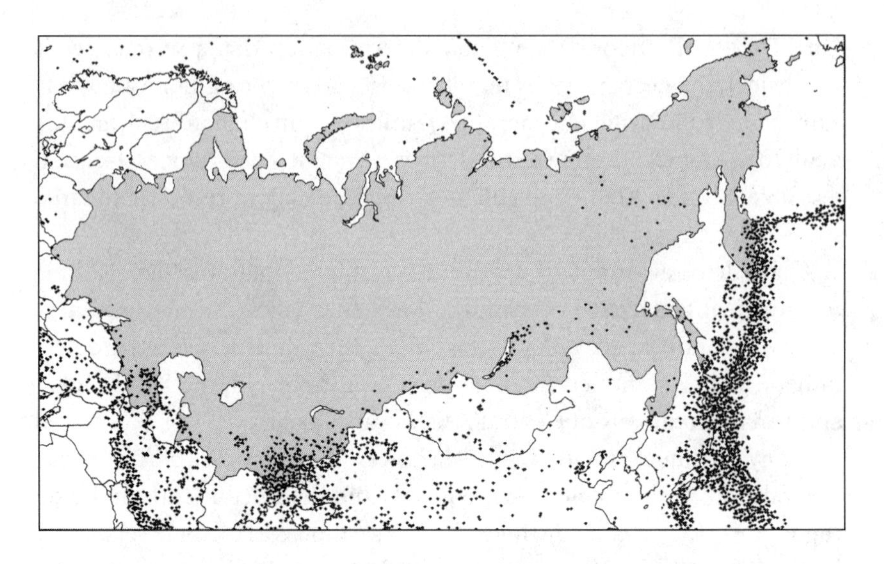

FIGURE 6.2

Earthquakes in the former Soviet Union and surrounding areas from 1971 to 1986.

Source: U.S. Geological Survey; Office of Technology Assessment, 1988.

Removing these and deeper shocks from consideration in 1963 still left a number of small shallow earthquakes in continental areas as difficult to identify (figure 6.2). With the breakup of the Soviet Union, however, many of the regions where these shallow events occur are now located in the independent countries of Armenia, Azerbaijan, Kazakhstan, Kirghizstan, Tajikistan, Turkmenistan, and Uzbekistan. The Russian Republic is less likely to test in those areas because secrecy would now be more difficult to maintain. In addition, much of the Russian Republic itself consists of old geological terrains, which contain few earthquakes.

FIRST MOTIONS OF SEISMIC P WAVES

The Conference of Experts meeting in Geneva in July and August 1958 thought that the first motions of seismic P waves could be used to identify earthquakes by observing clear downward, or dilatational, first motions of P waves. Nuclear explosions do not generate dilatational first motions, only compressions. It soon became clear in late 1958, however, that the first motions of short-period P waves were not reliable for events comparable in size to the U.S. underground explosions *Rainier* and *Blanca*.

With the stations of the World-Wide Standardized Seismograph Network (WWSSN) in place after 1963, clear first motions for earthquakes larger than magnitude m_b 5.5 often could be identified on long-period (low-frequency) records. First motions are either up or down on vertical instruments from earthquakes. Because first motions were not viable in 1963 to reliably identify events of m_b 4.75 and smaller, other identification methods clearly were needed for those events.

SEISMIC ARRAYS

Seismic arrays are groups of instruments typically spaced a few to about 6 miles (10 km) apart that are connected to one common recorder. Since 1963, additional seismic arrays have been installed around the world to detect and identify short-period seismic P waves. The signals from the group of instruments in an array can be processed to enhance a specific seismic signal and to suppress earth noise. This allows seismologists to identify more small events and to determine the direction of the source relative to the array.

Prior to the U.S. *Longshot* nuclear explosion in the Aleutians in 1965, the British group working on nuclear verification rushed to install new seismic arrays at Yellowknife, Canada, and in central Australia and India. These were in addition to the existing UK array in Scotland. Since 1996, when the CTBT was signed, several countries have installed many new arrays. The seismic wave called pP (see figure 3.1), which is reflected from the Earth's surface near the hypocenter and arrives after the P wave, can now be identified more readily using array processing and used to determine depths for many small to moderate shallow earthquakes.

DETECTION OF SEISMIC SURFACE WAVES FROM EXPLOSIONS

In 1958 Jack Oliver, my PhD advisor at Lamont, observed long-period seismic waves at Palisades, New York, generated by the 22-kiloton *Blanca* test. The long-period waves he observed were surface waves, not P or S waves. Over the next twenty years, his observation led to much study at Lamont and elsewhere on the use of surface waves to distinguish the signals of underground explosions from those of earthquakes.

When I arrived at Lamont in 1960, funding for the identification of underground nuclear explosions under the Vela program of the Defense Department had just started. I shared an office in Lamont Hall, originally the home of the Lamont family, with three scientists, including Paul Pomeroy, a graduate student who was several years ahead of me. For his PhD thesis, Pomeroy examined long-period recordings from nuclear explosions, mainly those in the atmosphere, as recorded at Palisades and at the long-period seismograph stations deployed by Lamont during the International Geophysical Year (IGY) in 1957 and 1958. In the early 1960s, Lamont and Cal Tech competed to discover methods for distinguishing the seismic signals of underground nuclear explosions from those of earthquakes.

USE OF SURFACE AND P WAVES FOR EVENT IDENTIFICATION

In 1963 Pomeroy and Lamont graduate student Robert Liebermann, now a retired professor at Stony Brook University, discovered a method called Ms-m_b that was to become one of the most reliable techniques to discriminate the signals of earthquakes from those of underground nuclear explosions. Magnitude Ms is a measure of long-period surface waves with periods of about twenty seconds. The period of a wave is the time it takes for one cycle of motion to occur—the time from one upward motion to the next—in this case, twenty seconds. Surface waves travel around the Earth's circumference, not through its very deep interior (figure 3.1). Magnitude m_b is determined from measurements of short-period P waves, which arrive before the surface waves, taking a shorter time to travel to a station through the deep interior of the Earth. The difference between Ms and m_b is then determined and used to identify an event as either an earthquake or an underground explosion.

This method takes advantage of the differing nature of the two types of seismic sources. Underground nuclear explosions instantaneously crush a relatively small volume of surrounding rock, which absorbs the force of the blast and generates seismic waves that propagate outward in all directions. In contrast, earthquakes are caused by slip (displacement) along a fault. Rupture in an earthquake takes place more slowly than in a nuclear explosion. Hence, earthquakes generate seismic waves for longer periods of time. They also rupture larger areas and radiate seismic waves whose amplitudes vary in azimuth (direction) around the source of an earthquake. This effect is much like that of a radio antenna that beams signals preferentially in certain directions. For earthquakes and underground explosions of the same m_b, earthquakes generated much larger surface waves. These measurements are shown in figure 6.3. British, Canadian. and other U.S. scientists went on to do a great deal of work using this technique.

In the Ms-m_b diagram in figure 6.4, when Ms is plotted on the vertical axis and m_b on the horizontal, a very good separation is obtained for earthquakes and underground explosions. One exception was a small

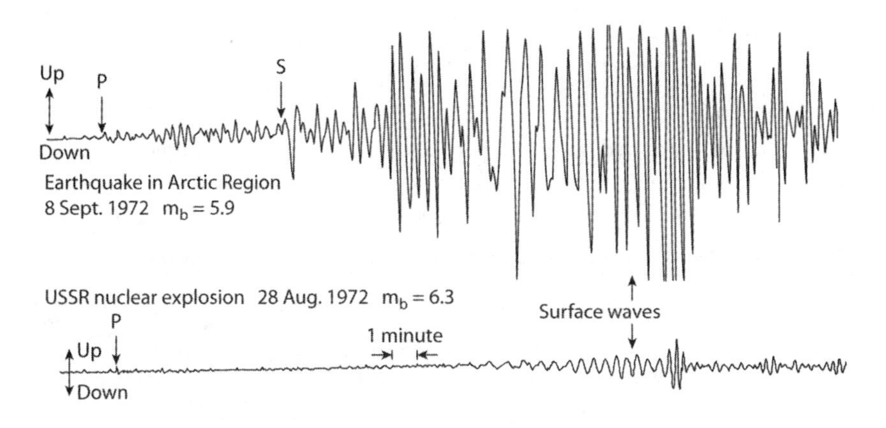

FIGURE 6.3

Seismograms of long-period waves from an earthquake in the Arctic near the USSR (top) and a Soviet underground nuclear explosion at Novaya Zemlya (bottom). Each was recorded at Eilat, Israel, at about the same distance from the two sources. The short-period P waves (not shown here) were of nearly the same magnitude, m_b, yet the surface waves are much larger for the earthquake.

Source: Sykes, Evernden, and Cifuentes, 1983.

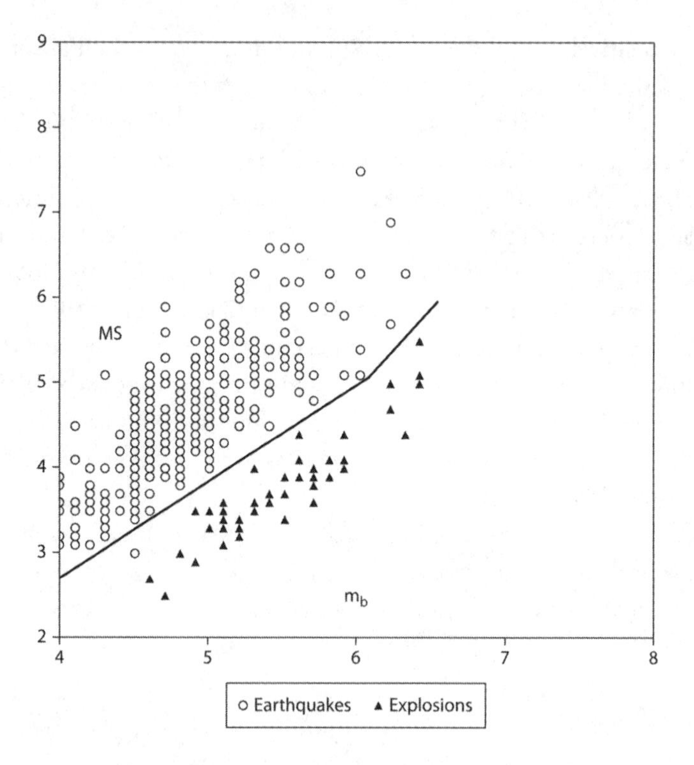

FIGURE 6.4

Magnitude Ms measured from long-period surface waves and magnitude m_b determined from short-period P waves. Note the clear separation of explosion and earthquake populations, with the exception of one earthquake in the southwest Pacific.

Source: Sykes, Evernden, and Cifuentes, 1983.

earthquake in the southwest Pacific where the network used to measure those two magnitudes was weak. The network was much more sensitive for measuring seismic events in the northern hemisphere. For a given yield, m_b readings are larger for earthquakes and explosions in regions of old strong crust and uppermost mantle of the Earth than they are for younger regions like Nevada. The values of m_b were corrected for this effect in figure 6.4.

Unfortunately, the development and validation of the Ms-m_b technique were a few years too late to be considered during the test ban negotiations in 1963. At the time, the U.S. government concluded that effective identification of underground explosions was too difficult. In addition, the Soviet Union was not willing to accept U.S. and UK requests for on-site inspections and monitoring stations on its territory. Consequently, underground tests were excluded from the Limited Test Ban Treaty (LTBT) between the United States, the USSR, and the UK, which outlawed explosions in the atmosphere, in space, and underwater, but not underground. Unfortunately, the Defense Department initially classified the Lamont work like that shown in figure 6.4 even though it provided a good international tool for event identification vital to a test ban.

ADVENT OF PLATE TECTONICS

Plate tectonic theory did not exist in 1963, but by 1969 it provided a new foundation for understanding why most earthquakes, volcanoes, and young mountain belts occur where they do. At the same time, it helped us to understand and predict the type (styles) of earthquake mechanisms for regions of the USSR and China, allowing for better discrimination between earthquakes and nuclear blasts. About half of my second book is devoted to my involvement with the plate tectonics revolution.

Plate tectonics also furnished an understanding of why seismic P waves propagate efficiently in the old rocks beneath the Soviet, Chinese, and Indian test sites but not beneath the U.S. test site in Nevada and the French site in southern Algeria. Until as late as 1988, the United States government failed to appreciate or to acknowledge this difference and accused the Soviet Union of cheating on the Threshold Test Ban Treaty.

During these years, many people in the U.S. departments of Defense and Energy either ignored or resisted seismological evidence that took these differences in the structure of the Earth into account. They also insisted that seismic surface waves should not be used for yield estimation or calibration. Debates over Soviet yields and verifying a Comprehensive Test Ban Treaty continued for many years. Throughout this time, I, along with other colleagues, fought hard to give voice to the notion that the United States was overestimating the yields of Soviet nuclear explosions by large amounts. I testified about this before Congress several times between 1972 and 1986.

DEVELOPMENT OF NEW SEISMOGRAPHS FOR NUCLEAR VERIFICATION

In the late 1960s, Paul Pomeroy, George Hade, and others at Lamont developed seismographs that could detect even longer-period (lower-frequency) seismic surface waves than those with periods of about twenty seconds that were then being used in the standard Ms-m_b method for distinguishing underground explosions from earthquakes. These new instruments, called High-Gain, Long-Period seismographs (HGLPs), magnified signals in the Earth more than one hundred thousand times for periods between twenty and seventy seconds—about one hundred times greater than the long-period instruments of the World-Wide Standardized Seismograph Network (WWSSN), which were installed in 1962 and 1963.

The HGLP instruments took advantage of the fact that natural earth noise drops off significantly for periods longer than twenty seconds and reaches a stable minimum at a period of thirty to forty seconds. Reduced earth noise at those longer periods enabled surface waves from even smaller earthquakes to be observed and used to distinguish signals from underground explosions from those of earthquakes.

Installing those seismometers in thick hemispherical steel tanks and deep underground in mines reduced earth noise, especially on horizontal-component HGLP instruments. One of the most important findings from the first HGLP station, in a deep mine at Ogdensburg, New Jersey, was that the discrimination of earthquakes and underground explosions was enhanced using forty-second surface waves compared to using the standard twenty-second waves. The period of a surface wave is the time it takes for one cycle of wave motion to occur—say, up to down and back up again on vertical records. It is not the length of time a surface wave lasts on a seismogram, which is many periods.

Recording quiet horizontal motions allowed another type of surface wave, called Love waves, to be detected. Love waves, like light, have horizontal vibrations perpendicular to the direction of wave propagation.

Named for applied mathematician A. E. H. Love, they could be employed when Rayleigh waves, the more common surface waves used to measure Ms, were not recorded well. Rayleigh waves can be unusually small for certain types of focal mechanisms of earthquakes that occur at depths of about 25 miles (40 km). The use of Love waves also provided an excellent separation between earthquakes and explosions on an Ms-m_b plot for periods of twenty to forty seconds.

The results from Ogdensburg in distinguishing the signals of explosions from those of earthquakes were spectacular enough that in 1970 the U.S. Defense Department funded the deployment of additional HGLP stations around the world set up by Lamont, the University of Michigan, and the Albuquerque office of the U.S. Geological Survey. The data from these stations were recorded digitally, a first for university seismologists. Many Lamont graduate students and staff worked on the installation of seven additional HGLP stations around the world and on the analysis of the data. Peter Ward, a postdoctoral student in seismology, headed the installation and operation of the Lamont HGLP stations. I was the principal investigator and was involved in analyzing the data to test their value for event discrimination.

The seven HGLP stations operated by Lamont showed similar very positive results for seismic discrimination to those at Ogdensburg. A minimum of earth noise was observed from all of these stations, which, along with the use of digital filtering, improved the detection levels for surface waves.

The HGLP stations operated in the 1970s at a time when the United States and the Soviet Union detonated a number of very large underground nuclear explosions, which the bilateral Threshold Test Ban Treaty forbade after March 31, 1976. (Today, modern digital recording stations around the world record seismic waves of similar long-period as well as shorter-period waves.)

John Savino, Peter Ward, and I of Lamont presented papers on the first results of the Ogdensburg HGLP experiment at the ARPA Conference on Nuclear Test Detection at Woods Hole, Massachusetts, in July 1970. Jack Evernden, then at the Advanced Research Project Agency (ARPA) of the U.S. Defense Department, organized the conference. In his written summary of the meeting, Evernden indicated that the CTBT detection and identification problem for underground explosions and earthquakes was solved. His summary and other papers presented at the conference were openly distributed in 1971. Evenden, however, failed to get his written summary cleared by Steven Lukasik, the director of ARPA. Lukasik exclaimed that the policy position in Evernden's summary was not approved by the Defense Department and demanded the return of all copies. Senator Clifford Case of New Jersey, a proponent of the CTBT, asked to see a copy of the entire document. When he found shreds of paper where the summary was torn out from the rest of the papers, he raised hell in the Senate.

JACK EVERNDEN AND CARL ROMNEY

This is a good time to discuss the major efforts Evernden made on behalf of the CTBT, as well as the role of Romney. I worked with Evernden on a number of occasions and published several papers with him on test ban verification. He was born in 1922 and received his PhD in geophysics from Berkeley in 1951. Along with physics professor John Reynolds, Evernden developed instrumentation for measuring and dating volcanic rocks younger than several million years. To date the time those rocks were formed (solidified), they made precise measurements of one isotope of the element potassium and of its radioactive decay product, an isotope of the element argon.

Allan Cox and his associates at Stanford University and the U.S. Geological Survey subsequently used the potassium-argon technique widely not only to date young volcanic rocks but also to ascertain whether those rocks had been weakly magnetized, either in the direction of the Earth's present field or, at times in the past, in the opposite orientation. In 1963 Fred Vine and Drummond Matthews of Cambridge University proposed that volcanic rocks of the seafloor, upon cooling, acquired this positive or negative magnetic signature or imprinting. In 1966 Lamont scientists Walter Pitman and James Heirtzler published their famous "magic magnetic anomaly profile" across the mid-ocean ridge in the east Pacific. These anomalies or departures from the Earth's magnetic field were generated as the seafloor cooled and was magnetized at different times when the magnetic field was either in its present direction or the opposite. It became a cornerstone for understanding that the Earth's outermost solid layer is composed of nearly rigid plates that move with respect to one another. Evernden also used the potassium-argon technique to date sediments containing bones of human ancestors (hominids) from the Rift Valley of East Africa in Kenya.

Evernden was so convinced of the immense importance of a Comprehensive Test Ban Treaty and the control of nuclear weapons that he relinquished his tenured full professorship at Berkeley in 1965 to work on methods of better nuclear verification, first at the Air Force Technical Application Center (AFTAC) and later at the Advanced Research Projects Agency (ARPA), both agencies of the Defense Department. He made a major sacrifice; if he had continued at Berkeley, he could have worked on what was to become plate tectonics. Evernden was a proponent of the HGLP seismic stations when he was at ARPA and personally worked with several of us on the analysis of those data at Lamont. Evernden did outstanding work on seismic verification at AFTAC and ARPA.

Many officials in the Department of Defense, however, were in favor of continued nuclear testing and the development of new nuclear weapons. An irony was that ARPA was responsible for research on nuclear verification and AFTAC for more routine classified monitoring of the Soviet Union and China.

Evernden's chief nemesis at AFTAC, and later at ARPA, was Carl Romney, who was born in 1924, two years after Evernden. Each received a PhD in seismology from Berkeley under seismologist Perry Byerly. Romney and Evernden were the only people at either AFTAC or ARPA who were experts in seismology. Evernden had three weaknesses. He was a poor listener, speaking for 95 percent of a conversation. He was argumentative and suffered neither those he thought were fools nor those who were poorly informed. Although he did excellent work on verification, those weaknesses hindered him in affecting U.S. policy.

In contrast, Romney, who grew up in Utah, was a good speaker and was very effective in influencing government policy, especially at the secret level. He was often the only seismologist involved in classified U.S. government committees concerned with monitoring nuclear testing and determining the yields of Soviet underground nuclear explosions. Many nonscientists in the government took Romney's opinions as gospel.

Romney stated in his book of 2009 that his own involvement with nuclear test detection began in mid-September 1949 after he had completed most of the requirements for a PhD at Berkeley. In 1955 he was hired by AFTAC, where he became assistant technical director for geophysics. In 1956, before finishing his PhD, he became the chief seismologist for the consulting firm Beers and Heroy, a contractor for classified research in seismology that was sponsored by AFTAC's predecessor, AFOAT-1. The firm had no seismologist until Romney joined them.

As the Apollo lunar program was winding down at Cape Kennedy, Florida, a powerful member of Congress convinced the Department of Defense to move AFTAC from Alexandria, Virginia, to Patrick Air Force Base near Cape Kennedy. Evernden told me that because Romney's wife did not want to move, Romney transferred to ARPA in 1972 so that they could stay near Washington, DC. He accepted a position in the Nuclear Monitoring Research Office of ARPA, where he became deputy director in 1980.

Improving seismic verification was the main goal of ARPA's Vela Uniform program, which funded research at universities in the 1960s and early 1970s. Unfortunately, few of the important findings at universities were utilized by AFTAC to improve routine monitoring and data analysis. Romney, as he recounted in 2009, had little use for many academics. After he moved to ARPA, he spent more funds on consulting companies to do routine data processing and far less on university research.

Few scientists outside of AFTAC were aware of the important data being collected from its classified seismic arrays around the world. Romney deserves credit for acquiring very quiet sites for those arrays, which produced high-quality data. These arrays or clusters of monitoring instruments, while extremely expensive, were well worth the investment. Evernden, who worked for Romney at AFTAC, tried unsuccessfully to get papers approved by the Defense Department to demonstrate how much progress had been made at the classified level in seismic monitoring of underground explosions.

7

CONGRESSIONAL HEARINGS ON A COMPREHENSIVE TEST BAN

From October 27 to 28, 1971, the Joint Committee on Atomic Energy of the U.S. Congress held hearings on the Status of Current Technology to Identify Seismic Events as Natural or Man Made. James Brune of the University of California at San Diego submitted a statement endorsed by a number of scientists knowledgeable about seismic monitoring regarding the verification of an underground test ban. They were all familiar with the results of the Ms-m_b method used for seismic identification as well as work presented by several of us at the Woods Hole conference in 1970.

The written statements of those scientists were published in the Hearings Before the Subcommittee on Arms Control, International Law and Organization of the Committee on Foreign Relations of the U.S. Senate—*Toward a Comprehensive Nuclear Test Ban Treaty*—on May 15, 1972. Their statements described recent advances that improved the detection of signals from underground nuclear explosions and earthquakes down to magnitude m_b 4 and yields of about two kilotons in hard rock. Those methods could be used to identify (differentiate) those events with high reliability upon deployment of adequate numbers of modern instruments. This, of course, was better than the threshold of m_b 4.75 proposed in 1963. This work was praised by members of the Committee on Foreign Relations but not by several members of the Joint Committee on Atomic Energy.

Lukasik, the director of ARPA, testified at the Joint Committee's hearings in October 1971 that we knew by then how to identify explosions down to a magnitude m_b of 4.5 (7 to 14 kilotons in hard rock) and that, in principle, seismic discrimination around m_b 4.0 (1 to 2 kt) appeared feasible. But he, like members of the Joint Committee, stressed that it would take time to verify such improvements.

I testified before the same subcommittee of the Senate Foreign Relations Committee on May 15, 1972, my first testimony before Congress. I was nervous. All of the other people who testified were seasoned, well-experienced witnesses and speakers. I was invited to testify about the results of the HGLP experiment and other advances in distinguishing the signals of underground explosions from those of earthquakes. Dr. Herbert Scoville, a test ban proponent who had worked at the CIA in the early 1960s and later at the U.S. Arms Control and Disarmament Agency (ACDA), recommended me to the subcommittee. I believe Evernden did as well.

My conclusions were very similar to those of the above group of scientists. We all stated that major advances in seismic identification had been made in the previous decade. Referring to the plate tectonic revolution of the late 1960s, I stated, "During the last ten years the earth sciences, which include seismology, have experienced a renaissance in the understanding of large-scale earth processes such as why earthquakes occur where they do, and what types of movements occur in various parts of the world."

Few seismologists outside of ARPA and AFTAC, including me, were aware at the time of just how good the U.S. classified capabilities were. The findings reported in my congressional testimony of 1972, as well as those of the scientists who wrote to Congress in 1971, were based on unclassified results. In addition, because we made no assumptions about future data that might be obtained from seismic stations within the USSR and China, we were of necessity discussing observations at distances greater than about 1850 miles (3000 km).

In my testimony, I also discussed the false alarm problem—that is, small earthquakes being misidentified as underground explosions. I stated that 99 percent of earthquakes in Eurasia (Europe plus Asia) of magnitude 4.5 and larger could be positively identified as such.

I addressed several possible evasion schemes, including detonating small nuclear explosions near the times of earthquakes, conducting multiple explosions so that their Ms value appeared to be that of an earthquake, and using decoupled (muffled) testing in large underground cavities. I stated that research on better processing of seismic signals might render the first two schemes less useful for evasive testing, which shortly turned out to be the case. I also pointed out that cavity decoupling—large explosions set off in deep underground chambers—had not been tested for a nuclear explosion larger than a fraction of a kiloton.

Hide-in-earthquake ceased to be a successful evasion method decades ago. Simple data processing methods allowed the signal from an explosion of interest to remain large while those from distant earthquakes were reduced in size. For example, the signal from a large earthquake in the eastern part of the Soviet Union hid that of the small explosion at the Soviets' Eastern Kazakhstan test site on the normal seismic recording (see upper part of figure 11.1), but playing out the recordings at a higher frequency clearly brought out the signal from the small explosion (see the lower part of that figure).

Detonating multiple nuclear explosions close in time to one another so that their signals were like that of an earthquake also is no longer a viable evasion scheme. Recording data over many periods and frequencies, as with the HGLP instruments, eliminated multiple explosions as a threat. While they might be made to look like an earthquake at one frequency, they could not at other frequencies.

Senator Edmund Muskie, the chair of the subcommittee, asked me only one question: "Are ARPA seismologists and private seismologists far apart in their assessment of the problem?" I answered, "I think not so far as detection and discrimination. I think the disagreement mainly resolves around the evasion issue." Muskie went on to include for the record the statement made by the group of scientists in April 1972. I think he may have expected me to launch into a criticism of the removal of the summary written by Evernden for the published volume on the Woods Hole conference, which I did not do. Muskie already knew about its removal.

When I returned to Lamont, I received a phone call from William Best of the Air Force Office of Scientific Research, who oversaw the HGLP project. He said that an official at ARPA, who was present at my testimony, said that I had done well, probably meaning I had not criticized ARPA. I found out later that person, a geologist, strongly believed that testing in huge cavities in salt could muffle large nuclear explosions.

I vowed to myself that in future testimony I would state clearly in my conclusions that a full test ban was verifiable, desirable, and in the national interest. Prior to the deletion of Evernden's summary for the Woods Hole conference report in 1971 and my Senate testimony of 1972, I was of the naive belief that advances in research on seismic monitoring soon would be translated politically into a full test ban.

In the subcommittee hearings of May 15, 1972, Philip Farley, acting director of the U.S. Arms Control and Disarmament Agency, presented the current views of the executive branch on the question of further limitations on nuclear tests. He said that he could confirm that the administration's position remains as stated by President Nixon in 1969: "The United States supports the conclusion of a comprehensive test ban, adequately verified." His statement gave no indication of what constituted adequate or sufficient verification.

Dr. Herbert York, dean of graduate studies at UC San Diego, head of the Livermore Lab from 1952 to 1958, and a top Pentagon scientist in the Eisenhower administration, also testified in May 1972. He prefaced his remarks at the 1972 hearing by quoting from an interview with the chairman of the Atomic Energy Commission, James Schlesinger, an opponent of a full test ban. Schlesinger said, "We have very much improved our capability to monitor events in the Soviet Union, but the question of the desirability of a test ban goes far beyond the question of monitoring what the Soviets are doing. The real question is, given the changed nature of the opposing forces in this era of negotiations, whether it is desirable to cease testing." York went on to say, "I agree with his position. I have for some time felt that the real question is whether it is desirable." York then summarized his testimony: "[W]e could have had and should have had a comprehensive test ban ten years ago." He discussed the main objections raised by others a decade before and found them largely wanting: (1) concern about a Soviet ABM (anti-ballistic missile) capability, (2) breakthroughs and surprises, (3) improvements in design factors such as yield-to-weight ratio of warheads on intercontinental missiles, and (4) our understanding of weapons effects. When asked about the prospects for a Comprehensive Test Ban Treaty (CTBT) in 1972 to halt development of multiple independently targetable reentry vehicles (MIRVs) for Soviet missiles, he said it was probably too late. At that time the Soviet Union likely was testing weapons for their MIRV land- and sea-based missiles. The United States had tested them earlier.

The CTBT hearings of 1971 and 1972 did not lead to negotiations for a full test ban treaty. The idea did not resurface in the U.S. government until 1977, when the Carter administration entered into negotiations with the Soviet Union and Britain. Following the Limited Test Ban Treaty (LTBT) of 1963, however, progress was made in other arms control negotiations, such as limits on the numbers of delivery vehicles for nuclear weapons and anti-ballistic missiles.

NUCLEAR NONPROLIFERATION TREATY

Four years of multinational negotiations finally led to the Nonproliferation Treaty (NPT) that was signed in 1968 and entered into force on March 5, 1970, for a duration of twenty-five years. It was the most important arms control agreement following the Limited Test Ban Treaty. The NPT was discriminatory by design, in that it divided countries into two groups: the acknowledged nuclear weapons states and nonnuclear countries. The weapons states pledged not to transfer nuclear weapons to other nations and not to assist any nation in making or acquiring them. The nonnuclear states agreed not to make or receive nuclear weapons and to allow the International Atomic Energy Agency (IAEA) to monitor their compliance. The parties to the NPT agreed to share the benefits of peaceful uses of nuclear energy.

The NPT requires parties to the treaty to "seek to achieve the discontinuance of all test explosions of nuclear weapons for all time and to continue negotiations to this end" and to "pursue negotiations in good faith on effective measures relating to the cessation of the nuclear arms race at an early date and to nuclear disarmament." Understandably, during each five-year review conference for the NPT, many of the nonnuclear states that signed the NPT sharply criticized the weapons states for not achieving a CTBT. A few countries, including India, Israel, and Pakistan, did not sign the NPT. North Korea signed but withdrew prior to its first nuclear test. India claimed that its first test, which was underground, was for peaceful purposes, which many doubted. India signed the Limited Test Ban Treaty, but neither the NPT nor the CTBT.

The NPT was up for renewal on its twenty-fifth anniversary in 1995. In return for extending it to infinite duration, the nonnuclear states extracted from the weapons states a pledge to reach a CTBT in 1996. The latter complied with their pledge. The NPT was extended indefinitely in 1995.

MONITORING AFTERSHOCKS OF A LARGE UNDERGROUND NUCLEAR EXPLOSION

A group of us from Lamont monitored aftershocks generated by *Benham*, one of the largest (1150 kt) underground nuclear explosions at the Nevada Test Site on December 19, 1968. Seismologists at the University of Nevada claimed that a previous large explosion at NTS had triggered earthquakes 25 miles (40 km) away. Their evidence was not very strong. Peter Molnar and Klaus Jacob drove a Lamont van, which had a small area for working and sleeping, to Nevada to monitor activity for a week before the United States detonated *Benham*. They operated six portable instruments in abandoned mines on the east side of the test site within 50 to 110 miles (75 to 175 km) of *Benham*.

I flew to Las Vegas on the day of the explosion and drove a rental car about 100 miles (160 km) to the town of Alamo, where Molnar and Jacob were staying. Halfway to Alamo, snow started to fall heavily, a weather condition that was supposed to be ruled out when deciding to fire such a large explosion. Those responsible at the test site probably wanted to complete the *Benham* test before the upcoming holidays. Fortunately, *Benham* did not vent—that is, leak radioactive materials into the atmosphere.

In a 1969 paper, Molnar, Jacob, and I found no significant change in the numbers of small earthquakes within 15 miles (25 km) of each instrument from before to after *Benham*. The explosion triggered numerous very small aftershocks, but they were confined to within about 6 miles (10 km) of its epicenter.

In late December 1968, after monitoring *Benham*, Jacob and I decided to move the portable Lamont instruments to Death Valley, California, just west of the Nevada border, to monitor several active faults. Molnar returned to Lamont. Our van was having mechanical problems, so we decided to stop midway for repairs in Las Vegas.

Because the number of working daylight hours was short in late December, we typically got up well before daylight for breakfast and ate dinner after dark. We went to a casino in Las Vegas near our motel for breakfast the next morning about 5 AM. TV sets had been set up so that patrons could see the splashdown in the Pacific of the Apollo flight that first flew around the moon but did not land on it. We were happy to see the successful landing while having breakfast, but most gamblers were oblivious to the exciting scenes on the TVs.

After we set up our instruments in Death Valley later that day, it was too cold to sleep in the van, so we drove back uphill to a motel in tiny Death Valley Junction, the same route on which mule teams hauled borax to trains at the Junction. Those trains and the mule teams had long disappeared. The next morning, December 31, at 5 AM, we went for breakfast at the only diner in town. A man came over to our table and remarked that we must not be from around there and was delighted to hear we were from New York. He told us that his wife, Marta Becket, would be dancing a ballet that evening, New Year's Eve, in town at the Amargosa Opera House and invited us to join him for the performance. It proved to be a memorable evening. The year before, as Marta and her husband were on their way from Los Angeles to Las Vegas, they passed a recreation hall in town. She decided to rent it, renaming it Amargosa, its original name.

Klaus indicated that he would be ten minutes or so late for the performance because he wanted to call his sister in Massachusetts to wish her Happy New Year. Becket's husband said, "No problem; we'll wait for you to start." We turned out to be the only people who stayed for the full performance. One family came late and left early. Becket's husband pulled the curtain, worked the lights, and played recorded music. On the back wall of the small theater, he and Marta had painted a copy of Velasquez's *Las Meninas*, in which the young infanta Margaret Theresa is surrounded by her entourage of maids of honor, a chaperone, a bodyguard, two dwarfs, and a dog. They said they went ahead with ballet performances even if no one else came; they had *Las Meninas* for spectators. The Opera House, now on the National Register of Historic Places, was discovered several years later and became successful. Marta's final show was in 2012.

Afterward, Klaus and I were invited to have a glass of champagne with Marta and her husband. About an hour before midnight, as we stood around celebrating, a frightened woman's face appeared at a window of their home. She said her husband, who was to get out of prison at midnight, had vowed to kill a man, and she needed help in finding him. As Ms. Becket's husband was about to join her, Marta implored him not to go, but he said, "These are our neighbors, and I need to help them." So off he went. We headed back to our motel room and slept soundly as the New Year arrived.

8

PEACEFUL NUCLEAR EXPLOSIONS

During the Moscow negotiations for the Threshold Test Ban Treaty (TTBT) in 1974, the Soviet delegation insisted on two major conditions. One condition, as mentioned earlier and discussed in more detail in later chapters, was that the threshold be based on yield, not seismic magnitude. A second was that the treaty be limited to the testing of nuclear *weapons* underground and not apply to peaceful nuclear explosions (PNEs). The United States accepted both conditions, and the two parties signed the TTBT in July 1974. Soon after, negotiations were started on a separate bilateral treaty dealing with peaceful nuclear explosions; it was signed two years later in 1976.

The United States and the Soviet Union used peaceful explosions to create large underground cavities in salt for storage of various products, to break rock for petroleum recovery, and as large sources of energy for seismological studies of the Earth's crust and upper mantle. The numerous cavities created in salt by peaceful explosions have been of great concern because some could be used for evasive (muffled) nuclear testing. One large cavity at Azgir in western Kazakhstan was, in fact, used by the Soviets for a partially decoupled test. It was the largest evasive test ever conducted.

In 1974 the Soviet delegation stated that their country needed to conduct numerous peaceful explosions to benefit its national economy. A major problem, however, is that a peaceful explosion cannot be distinguished from a weapon test by either seismic means or satellite imagery. For example, India claimed that its first nuclear test in 1974, which produced seismic waves like those from a weapon test, was a peaceful nuclear explosion. Years later it became clear that it actually was part of India's development of nuclear weapons. From this, the United States concluded that peaceful nuclear explosions needed to be subject to a treaty that would spell out what was acceptable.

Distinguishing weapons tests from peaceful nuclear explosions ultimately was resolved in the Peaceful Nuclear Explosions Treaty (PNET) of 1976. It stated that any nuclear explosion at the declared weapons test sites of the two countries would be considered a weapons test and would be subject to the terms of the TTBT of 1974. All nuclear explosions outside of the declared nuclear weapons test sites would be considered to be peaceful explosions and would be covered by the terms of the PNET.

THE U.S. PNE PROGRAM

The United States began peaceful nuclear explosions in 1957 in a program called Plowshare, which ended in 1973, a year before the Threshold Treaty was signed. Plowshare's two main purposes were excavating harbors and stimulating the flow of petroleum. In the 1950s and 1960s, many people in the United States, especially in the Atomic Energy Commission and the weapons laboratories, thought that PNEs could be utilized to accomplish many unique and important purposes. Others believed they were simply an excuse for continued nuclear testing.

Donald Springer of Livermore and others published a list of twenty-three peaceful nuclear explosive devices that were tested at the Nevada Test Site (NTS) from 1962 through 1971. Yields of all of these U.S. tests were declassified. One of the largest was the 104-kiloton *Sedan* explosion in 1962. That PNE excavated a huge hole with a diameter of 1280 feet (390 meters)—the length of more than three American football fields—and a depth of 320 feet (98 meters). The purpose of *Sedan* was to demonstrate that large holes could be created in the ground for various purposes using PNEs. Satellite imagery easily detected the crater it formed.

Some U.S. proponents advocated using very large nuclear explosions to create harbors in Alaska and Australia as well as megaton-size explosions to create a sea level Panama Canal during the time the United States controlled the Canal Zone. Fortunately, the United States did not go ahead with any of those projects because immense amounts of fallout would have resulted from those huge explosions. It is not difficult to imagine the great international political storm that would have occurred if that project in Panama had gone forward.

The United States conducted the *Gnome* explosion in salt in southeastern New Mexico in 1961. *Gnome* and the 1964 explosion *Salmon* in Mississippi, which was not a PNE, left underground cavities standing long afterward. Most cavities in materials other than salt collapse within seconds to days. The Soviet Union used nuclear explosions to create many cavities in salt for the storage of gas, oil, and waste products.

The United States conducted a number of nuclear explosions to obtain gas that was not accessible by standard drilling methods in tightly locked pores of very fine-grained rock. *Gasbuggy* of 29 kilotons was detonated in New Mexico in 1967, and 40-kiloton *Rulison* in western Colorado in 1969. *Rio Blanco*, set off in northwestern Colorado in 1973, consisted of three 33-kiloton explosions in a single hole but at different depths. Not very effective in stimulating gas, it was the last of the U.S. peaceful nuclear explosions.

I visited the *Rio Blanco* site about a year or two after it was detonated as part of a field trip for a U.S.-Japan conference on earthquake prediction in Salt Lake City. The Department of Energy (DOE) showed us a film about the project that included small children bouncing on the devices prior to their being lowered down a hole in an effort to demonstrate how safe PNEs were. I found the demonstration especially objectionable because several of our Japanese colleagues had lived through World War II and the bombing of Hiroshima.

There were several other objections to the use of PNEs to obtain gas. One was that the gas produced would be radioactive. Another was that a large amount of water would be needed in dry regions for further exploitation. Today, tight gas formations are being exploited in various parts of the eastern and central United States using hydrofracturing (fracking). Recent technology permits a horizontal well to be drilled from a vertical hole at depth into a tight geological formation such as shale. Fluids injected at high pressure along that horizontal hole create many vertical hydrofractures, which release trapped gas. The process is hotly debated today because it requires great amounts of water as well as the disposal of the toxic liquid wastes produced during fracking.

SOVIET PNE PROGRAM

The Soviet program of peaceful nuclear explosions (figure 8.1) started in 1965, later than in the United States, but it continued much longer, until 1988. The USSR conducted its first peaceful nuclear explosion of 125 kilotons at its Eastern Kazakhstan test site, where it formed a crater that is easily seen on unclassified satellite images. Yields of the explosion and some later PNEs were released by the Soviet Union at international conferences on peaceful explosions. D. D. Sultanov and others published locations, depths, times, names, and yields of 122 Soviet PNEs in 1999.

FIGURE 8.1

Sites of peaceful nuclear explosions in the Soviet Union.

Source: Bulatov, 1993.

During the 1974 negotiations for the TTBT, the Soviet delegation stated their country would like to use very large nuclear explosions to construct a major canal that would bring abundant water that normally flows into the Arctic Ocean through the Pechora River into the south-flowing Kama River, which empties into the water-starved Caspian Sea. A few nuclear explosions with yields up to about 100 kilotons were detonated along the line of the proposed canal in the early 1970s to test the concept.

The bilateral Peaceful Nuclear Explosion Treaty (PNET) of 1976 permitted the Soviet Union to perform a series of nuclear explosions whose total yield was larger than the 150-kiloton limit of the TTBT provided the United States could conduct extensive on-site observations. Each explosion in the series was to be no larger than 150 kilotons. On the advice of his science adviser, however, Secretary-General Gorbachev later canceled the Pechora-Kama project on the grounds that the proposed nuclear explosives would create dangerous radioactive fallout. After the PNET was signed in 1976, Russia did not detonate peaceful nuclear explosions nearly as large as 150 kilotons.

Soviet explosions in salt are important because the deep cavities that remain after those events could be used at some point to conduct decoupled or muffled nuclear tests. I made a major finding in 1996, discovering that none of the cavities created by the Soviets, with the exception of those at Azgir in western Kazakhstan, was large enough to be used to conduct a fully decoupled nuclear explosion with a yield larger than one kiloton.

It is important to note that Kazakhstan became a separate nation following the dissolution of the Soviet Union in 1991. The ninth largest country in the world in area, bigger than Western Europe, Kazakhstan has not detonated any nuclear weapons since its independence. It is very antinuclear as a result of fallout from past nuclear tests. Kazakhstan has ratified both the Nonproliferation Treaty and the Comprehensive Nuclear Test Ban Treaty.

Cavities created in salt by well-coupled nuclear explosions are only suitable for conducting fully decoupled nuclear explosions of much smaller yield, up to 5 percent as large as that of the explosion that created the cavity. Therefore, because of the limited size of cavities created in salt by past well-coupled nuclear explosions in the Russian Republic, they would be suitable for fully decoupled tests only of sub-kiloton size.

The locations of past nuclear explosions in salt in Russia and Kazakhstan are known very accurately. In addition, earthquakes in and near salt deposits are very few, so they can be recognized readily. Explosions in salt also are relatively easy to detect because salt transmits seismic waves as easily as hard rocks. Therefore, monitoring sites of previous nuclear explosions in salt in the Russian Republic is relatively easy.

NUCLEAR EXPLOSIONS IN SALT AT AZGIR

In 1966 the Soviet Union conducted a PNE of 1.1 kilotons in salt to the north of the Caspian Sea near the small town of Azgir, Kazakhstan. It formed a collapsed crater at the surface because the explosion was of quite shallow depth. The crater is visible on unclassified images made by the French SPOT satellite. In 1996 I made a special study of it, searching all available records from the worldwide seismic station network and Canadian stations for P waves from the event. Using sixteen stations, I obtained an average m_b of 4.5, a sizable magnitude for an event whose yield was only 1.1 kiloton. The ability to detect such a small event using data mainly from simple unclassified stations outside the Soviet Union reflects the efficient transmission of energy from underground explosions in salt at Azgir. Remember from earlier chapters that the seismic magnitudes of underground explosions at the Nevada Test Site with yields of about one kiloton were much smaller than m_b 4.5.

I obtained similar results for fourteen larger nuclear explosions in salt at Azgir between 1968 and 1979. Situated in a remote area, Azgir appears to have been a testing area for using nuclear explosions either to form cavities in salt or to conduct nuclear explosions within two of those cavities. Azgir is situated in the Pre-Caspian depression, which contains the largest concentration of salt domes in the world.

The Soviets detonated a second peaceful nuclear explosion north of Azgir in 1968. Its yield was reported as 25 to 27 kilotons. Soviet workers stated that its cavity filled with water after the 1968 shot. They used the cavity to set off six additional very small nuclear explosions with yields ranging from 0.01 to 0.5 kilotons (10 to 500 tons). The sensitive NORSAR seismic array in southern Norway recorded the four smallest events. Data from the two largest explosions in the cavity during 1975 and 1979 aided them in analyzing the smaller events. Their occurrence on nearly an exact hour early in the morning helped in their identification. Detecting the smallest nuclear explosion in that cavity with a yield of about 10 tons and a magnitude of 2.8 was a significant accomplishment. They would be even more easily detected today.

Soviet scientists stated that the small explosions in the water-filled cavity were intended to test their use for deep seismic sounding of the crust and upper mantle at large distances across their country. They were very well coupled explosions, not decoupled or muffled, because they were fired in water, making them valuable studies of the crust and mantle. Soviet scientists reported that one of those small explosions was used to generate elements beyond uranium in the periodic table.

The Soviet Union went on to conduct many nuclear and chemical explosions to explore its deep Earth structure along lines that extended thousands of miles (kilometers) (Figure 8.1). It probably used more PNEs for this purpose than for any other task. The seismic signals from PNEs of about 10 kilotons could be observed at those large distances, permitting Earth structures to be imaged as deep as 200 kilometers. Conducting such a large number of very large chemical explosions would have been more difficult.

Eight large underground nuclear explosions were conducted to the northeast of the small town of Azgir from 1971 to 1979. Those sites, along with numerous roads leading to them, were visible on unclassified SPOT satellite images taken in 1988.

The cavity created by the nuclear explosion of 64 kilotons in 1971 was used to conduct a partially decoupled nuclear explosion of 8 to 10 kilotons on March 26, 1976. It is the only known nuclear explosion that was significantly decoupled or muffled with a yield larger than that of the U.S. *Sterling* explosion of 0.38 kilotons.

OTHER SOVIET NUCLEAR EXPLOSIONS IN SALT

Sultanov and others reported that fifteen cavities were formed by PNEs with yields of about 3 to 8.5 kilotons from 1980 to 1984 at depths of 3000 to

3600 feet (900 to 1100 m) at Astrakhan near the mouth of the Volga River in the Russian Republic. All were intended for the storage of propane and butane within a large gas field. In 2001 William Leith of the U.S. Geological Survey wrote to me that most, and perhaps all, of those cavities had completely collapsed within several years of the explosions that created them. Those cavities were not used and remained air filled for years. That, coupled with the fact that salt at the depths of the Astrakhan explosions likely is not homogeneous and probably contains layers of other sedimentary rocks, contributed to the collapses.

The Soviet Union conducted several other PNEs in salt during the 1970s and 1980s just to the north of western Kazakhstan at Karachaganak and Orenburg within what is now the Russian Republic. A large area of layered salt deposits is well known to the north of Lake Baikal in Russian Siberia. The Soviet Union conducted a few small PNEs in that region as well. In 1996 I found that only one cavity, if it remains standing, might be suitable for fully decoupled tests as large as 0.5 to 0.9 kiloton.

Although peaceful nuclear explosions figured largely in debates about the Limited Test Ban Treaty in 1963 and the Threshold Test Ban Treaty (TTBT) of 1974, neither Russia nor the United States has conducted PNEs for decades. Data from past PNEs, however, are valuable in efforts to monitor the full test ban treaty.

Gorbachev and Bush finally signed a revised TTBT protocol and its companion Peaceful Nuclear Explosion Treaty (PNET). Those bilateral treaties were ratified and then entered into force on December 11, 1990.

9

HEATED CONTROVERSIES OVER YIELDS OF SOVIET TESTS AND AN UNSUCCESSFUL ATTEMPT AT A CTBT

In this chapter I examine the verification of the Threshold Test Ban Treaty (TTBT) during the Nixon, Ford, and Carter presidencies. Considerable controversy arose over the yields of Soviet weapons tests, much of it sad and frustrating, between 1974 and 1990. This controversy became known as the "yield wars."

A number of arms control advocates thought the United States should not have entered negotiations for a TTBT in 1974, believing that the testing threshold of 150 kilotons was too high. They much preferred either a full ban on underground testing or a much lower threshold for allowed testing of nuclear weapons. Nevertheless, neither of those was likely to have been attempted as the Nixon administration was coming apart during the Watergate scandal.

I became involved in the TTBT negotiations in 1974 on a moment's notice when I was suddenly asked to take part. I was motivated by the desire to lend my expertise in the field of seismology to do whatever I could to help bring about a treaty. Little did I know that the TTBT, signed in 1974, would take until 1990 to be renegotiated and finally to enter into force. The long debate about yields, with U.S. claims that the Soviets were cheating, postponed consideration of a full test ban treaty, or CTBT.

For decades, however, the United States had been using a method that greatly overestimated the yields of Soviet nuclear explosions, and it continued to do so for an additional fifteen years after the treaty was signed in 1974. The political consequences were huge, with the U.S. government incorrectly accusing the Soviet Union of cheating on the TTBT by testing well above its threshold. This occurred at a tense time during the Cold War.

The United States and the USSR each agreed that they would not test above the 150-kiloton limit of the TTBT after March 31, 1976, preventing the testing of megaton nuclear explosions by the two countries thereafter. It occurred too late, however, to prevent the testing of weapons carried by missiles with multiple warheads. In the twenty-one months between the signing of the TTBT in July 1974 and its start date in 1976, the Soviet Union and the United States each conducted a number of weapons tests much larger than the 150-kiloton limit of the treaty. They did not do so afterwards.

Each of the signatories to the TTBT was required to specify its sites for testing nuclear weapons. The United States chose the Nevada Test Site (NTS), and the USSR designated Novaya Zemlya and Eastern Kazakhstan. The treaty called for an exchange of information on the geological and physical properties of each geophysically distinct part of each test site. It also called for exchange of information on yields of two past large explosions within each distinct subarea. Those data were to permit better calibration of past and future nuclear explosions. The exchange of data was to occur upon ratification of the treaty, but this did not happen until 1990.

Contamination of the Eastern Kazakhstan site, the spread of radioactive debris off-site, and reports of widespread cancers led to demonstrations in Kazakhstan over continued nuclear testing. It proved to be a major factor in the breakup of the Soviet Union, as Kazakhstan became an independent country and testing was halted there in October 1989. The Russian Republic conducted tests until 1990 at its Arctic site on Novaya Zemlya.

Having received a secret security clearance from the U.S. government in 1968, I was one of a group of seismologists who were briefed on the determination of seismic magnitudes and yields of Soviet underground explosions using the classified capabilities of the Air Force Technical Applications Center (AFTAC) several months prior to the TTBT negotiations in 1974. Most of the discussion at that meeting, however, was about seismic magnitudes, including Soviet instruments, and why the Soviet Union obtained different magnitudes than the United States for either the same events or the same areas. The classified U.S. magnitude-yield curves were shown to our group but not discussed. That was a mistake, because yield determination was central to ascertaining if the USSR had exceeded the 150-kiloton threshold of the TTBT after March 1976.

Carl Romney of the Defense Department and Eugene Herrin of Southern Methodist University were among the few seismologists in the United States who long had access to and used classified determinations of magnitudes and yields of U.S. explosions. Romney knew that explosions of a given yield in hard rock and salt generated higher seismic amplitudes than those set off in soft rock. Romney mentioned in his 2009 book that early 1960s U.S. tests in Nevada in dry alluvium generated seismic waves (and their associated magnitudes, m_b) that were much smaller than those for tests in tuff, a relatively soft volcanic rock. The few U.S. explosions in hard rock in Nevada in the 1960s had generated the largest seismic waves (when differences in yield were taken into account). The *Gnome* and *Salmon* explosions in salt in 1961 and 1964 generated seismic waves that were even larger at stations in the central and eastern United States.

Nevertheless, Romney and others in the U.S. government insisted on setting the treaty threshold by magnitude rather than yield. Soviet scientists and officials undoubtedly knew that their numerous nuclear tests at sites in hard rock had generated larger seismic waves for a given yield; they would be at a distinct disadvantage if the threshold were based on magnitude and the United States were permitted to test in alluvium and tuff. Hence, a magnitude threshold was not acceptable to the Soviet Union.

EFFORTS TO DETERMINE SOVIET YIELDS: THE AFTAC PANEL

Once the TTBT was signed in July 1974, AFTAC made a major effort to pull together U.S. seismic data on yield determinations for Soviet and U.S. underground explosions. Thomas Eisenhauer and Robert Zavadil, senior scientists at AFTAC, studied surface waves from Soviet underground tests. They concluded in a classified document that Soviet yields calculated from surface waves were systematically smaller than those determined from short-period P waves (and the magnitudes m_b determined from them).

The United States used P waves, not surface waves, to calculate yields of Soviet explosions.

I determined to better understand the U.S. classified method of yield estimation, making it a top priority when I returned from the TTBT negotiations in 1974. Soon thereafter I became a member of a classified panel advising AFTAC on the seismic determination of yields of Soviet underground nuclear explosions. Except for its last meeting in California, the panel convened at Patrick Air Force Base in Florida. This was the first of a number of closely controlled panel meetings.

Chaired by Herrin, the panel convened several times until AFTAC asked that it no longer meet in 1977. Although it was not dissolved, no reason was ever given for why it never met again. Presentations filled the time allotted to the panel, but there was seldom time to discuss major conclusions. Interestingly, Herrin never sent a classified draft summary to each of us for comments. Instead, he either sent his own version directly to officials in the Defense Department or briefed them orally.

The panel first heard the results on yields determined from surface waves by Eisenhauer and Zavadil. Herrin and Romney argued vociferously that surface wave determinations of yield should neither be trusted nor used for yield determination because those waves were contaminated by "tectonic release": underground explosions in hard rock often trigger the relief of varying amounts of natural stress (pressures) in rocks near a shot point. Its effect was like adding the signal from an earthquake to that of the explosion itself.

For some explosions in Eastern Kazakhstan tectonic release was large, but for others it was quite small. I argued that those events with small tectonic release could be used to calibrate the magnitude-yield curve for P waves, but AFTAC refused to do so because Romney and Herrin were adamant that their formula using P-wave magnitudes, m_b, was accurate.

At another panel meeting, the commanding general of the base made an opening statement that we should stick strictly to the written agenda, which I think was put together by either Herrin or Romney or both. Under their powerful influence, that meeting and others had no effect on changing the U.S. procedure for yield estimation. The panel was usually asked to comment on seismic absorption and possible differences in magnitudes (called m_b bias) of the Eastern Kazakhstan test site with respect to Nevada. It was never asked, however, to comment on or even discuss larger questions, such as what was the best method of determining Soviet yields or whether the USSR had tested above the 150-kiloton limit of the Threshold Treaty.

The panel examined the existing procedure for yield determination using short-period P waves. It was necessary to use known data for explosions in hard rocks because most Soviet explosions at their two main test sites were situated in those rocks. The catch was that few U.S. explosions were detonated in hard rock. Data existed for only three early underground explosions in hard rock in Nevada—*Hard Hat* (5.7 kt), *Shoal* (12 kt), and *Pile Driver* (62 kt)—as well as magnitudes and yields of two French underground nuclear explosions in southern Algeria in 1963 and 1965. Several of us on the panel thought the seismic data for those U.S. explosions were poor. In addition, the yields, especially those for *Hard Hat* and *Shoal*, were much smaller than the 150-kiloton limit of the TTBT. The published yields of the two largest French explosions in granite—*Ruby* and *Saphir*—were 52 and 120 kilotons.

The Soviets had released the yield of only one nuclear explosion at their two main test sites: a 125-kiloton peaceful explosion that created a large crater at the Eastern Kazakhstan test site on January 15, 1965. It was difficult to use it for accurate calibration of the yields of contained underground explosions—that is, those that did not blow out at the surface as the event in 1965 did. The United States had released much more information on yields of its nuclear explosions. What was clear to me in 1974 and later was that there was no "magic bullet" for estimating Soviet yields. P waves had one set of problems and surface waves another. Our strategy should have been to come up with procedures that gave what we thought were the best determinations of yields and of their uncertainties. That did not happen.

Herrin and Romney, who were well aware that different magnitude-yield relationships applied to soft rocks like tuff and alluvium, were the principal architects of the m_b-yield curve used to calculate yields of Soviet tests in hard rocks. I think they strongly believed that the m_b-yield relationship for hard rock applied everywhere and that m_b for a given yield was only a function of the type of rock at the site of the explosion (the shot point). That premise, which turned out to be false, carried major scientific and political consequences until the issue finally was resolved fifteen years later.

As discussed in an earlier chapter, much evidence existed in 1974 that the magnitude m_b of explosions in Nevada and western Colorado varied in a consistent manner for paths to most seismic stations in the United States and Canada. For paths to stations in eastern and central North America, arrival times were early, indicating faster velocities, and magnitudes were larger, indicating smaller P-wave absorption.

DIFFERENT VIEWS ON YIELD DETERMINATION

Peter Marshall, a British scientist who worked on nuclear test verification, and two scientists from Livermore concluded that those findings for stations in North America likely were associated with differences in the Earth's upper mantle at depths of about 30 to 125 miles (50 to 200 km) at both the down-going and up-going ends of the paths traversed by P waves. In 1976 Marshall and Springer correlated those differences in travel times and absorption with differences in the speed of seismic P waves, called P_n, that travel in the uppermost mantle of the Earth just below the crust. P_n velocities average about 8.2 kilometers per second (km/s) in the central and eastern United States and about 7.85 km/s in the west.

Marshall and Springer found that differences in average P_n velocities correlated with differences (or residuals) in m_b values that averaged about 0.3 of a magnitude. This translates into a variation in yield by a factor of 2.5. Measurements of P_n were widely available for parts of various continents, including the USSR. Thus, they concluded that reliable measurements of P_n could be used to infer differences in the absorption of P waves in the upper mantle beneath various test sites. Hence, m_b values for Soviet tests could be corrected to better estimate their yields. Marshall and Springer mentioned that more accurate yields of Soviet tests were needed once the United States had signed the Threshold Test Ban Treaty.

It was well known before the advent of plate tectonics in 1967 that the upper mantle beneath Nevada and western Colorado is characterized by slow P_n wave speeds and greater absorption of short-period seismic P waves. These properties clearly differed between young areas like

Nevada and regions of older hard rocks, such as those in central and eastern North America and much of the USSR.

In 1979 Marshall of the UK and Springer and Howard Rodean of Livermore went on to quantify Marshall and Springer's earlier results for seismic stations and P_n velocities globally. For a given yield, they deduced that m_b magnitudes for hard rocks in Kazakhstan were about 0.38 magnitude units larger than tests in hard rock at the Nevada Test Site, which translates into a variation in yield by a factor of about 3.2. Their results for Kazakhstan were not at the Soviet test site itself but nearby. Kazakhstan, including the test site, contains large areas of very old hard rocks.

Soviet bulletins published the arrival times of P waves at their seismic stations. The station at Semipalatinsk near the Kazakhstan test site reported arrivals from earthquakes worldwide that were systematically early, like the early arrivals at stations in central and eastern North America. Early arrivals are indicative of high seismic velocities beneath stations.

Norwegian seismologists observed that P waves from Soviet tests recorded by their NORSAR seismic array, located on hard rocks, were rich in frequencies up to at least 8 cycles per second (8 Hz), whereas P waves from NTS explosions recorded in Norway lacked such high frequencies. This was additional evidence that absorption (attenuation) of P waves was low in the upper mantle beneath Soviet test sites and high beneath the Nevada Test Site.

The difference in m_b between Nevada and Soviet test sites for explosions of the same yield became known as the magnitude bias (m_b bias). It was a systematic effect, and not taking it into account resulted in large overestimates of the yields of Soviet tests. This difference was not trivial: U.S. estimates of Soviet yields were about three times too large. Arguments about magnitude bias and Soviet yields continued until 1988, when they were finally resolved.

At what turned out to be the last meeting of the AFTAC panel I was on in 1977, at least three of us—Thomas McEvilly of UC Berkeley, Springer of Livermore, and I—argued that the U.S. procedure for estimating Soviet yields from P-wave magnitudes (m_b) was incorrect and that surface waves needed to be included in some way. The magnitude m_b could be used, but it needed to be calibrated for differences in the absorption of P waves—that is, for magnitude bias.

Several of us, including me, asked for a straw vote of the panel concerning these important questions. A majority agreed with us. Herrin and Romney, however, disagreed. They held, and continued to argue for nearly another fifteen years, that it was premature to change the existing procedure for yield determination using P waves. I think the reason our AFTAC panel never met again lies in the strong opinions of Romney and Herrin, who refused to consider other scientific voices. A different AFTAC panel was formed years later to advise again on seismic issues, which Herrin again chaired for many years. Several of us from the original panel were not included. Herrin died in 2010.

AD HOC PANEL ON YIELD DETERMINATION UNDER THE DEFENSE SCIENCE BOARD

Following the last meeting of the original AFTAC panel, the U.S. Defense Science Board formed an ad hoc group on Soviet yields in 1977. A member of a consulting firm in northern Virginia, who had no expertise in either seismology or other areas of geophysics, chaired the panel. His firm, which had beautiful views of the Pentagon and the Potomac River from its offices in Roslyn, Virginia, did considerable classified work for the Defense Department. Clearly Romney instigated the formation of the panel, likely picked its members, and made the major presentation. He and I were the only members from the then defunct AFTAC panel. Romney may have asked that I be a member of the committee so he could claim that dissenting views on yield determination, such as mine, had been heard.

Other members of that panel included seismologists Carl Kisslinger of the University of Colorado and Donald Helmberger of Cal Tech, rock mechanist Eugene Simmons of MIT, a member of the staff at either the Sandia Weapons Lab or Los Alamos, and Richard Wagner of Livermore. None of them was familiar with yield determination. Helmberger was the only one who had worked on the seismic absorption of P waves. Wagner later became an influential hawk as an assistant secretary of defense during the Reagan administration. When he moved to the Department of Defense, he argued that the Soviets were cheating on the Threshold Test Ban Treaty and other arms control agreements.

Romney never mentioned arguments made by many of us on the recent AFTAC panel against using the Romney-Herrin relationship for yield determination. He mentioned neither the determination of Soviet yields using surface waves nor corrections to the seismic magnitude m_b based on differences in P_n velocities. He showed data from a single station in Missouri, where seismic waves were small like those in the western United States, but not data from the large numbers of stations in North America where just the opposite was observed. It was a very biased selection by Romney; it did not occur by chance.

Again, the agenda was set and tightly controlled. I was given about five minutes to counter Romney's claims and to mention the findings of the previous AFTAC panel. Helmberger was not helpful in stating that the magnitude bias between Nevada and Eastern Kazakhstan likely was small. Simmons unfortunately accepted Romney's assertions because he regarded him as an expert who worked full-time on seismic verification.

Romney, I discovered with time, was a very cagey person who knew his audience very well, including what he could get away with. He would give one story to a knowledgeable seismologist and another to people, like the others on this ad hoc panel, who were not familiar with what had gone on at the last AFTAC panel or what Marshall and Springer had concluded in 1976. I found Romney's presentation to the ad hoc panel outrageously deceptive. I never trusted him again. I think he likely regarded me as an enemy for my views on this panel and the AFTAC one that preceded it.

Romney persuaded the ad hoc group to endorse a one-page secret document stating that the Russians could be breaking the treaty by testing weapons with yields much larger than the 150-kiloton limit of the TTBT. How much greater is likely still classified. I came away very dejected with that outcome and angry about Romney's manipulation of the panel. That classified document "rang a lot of bells" in the U.S. government. At the time that yields of tests at Eastern Kazakhstan were being overestimated by many times, Romney concluded that the actual yields were even higher, not smaller.

REVIEW BY CARTER'S SCIENCE ADVISER

In 1977 Jimmy Carter was inaugurated as president, and seismologist Frank Press became his science adviser and head of the Office of Science and Technology Policy. I wrote an unclassified letter to Frank, whom I had known for a long time, saying I had been part of the AFTAC panel and the recent ad hoc committee. I stated that I strongly disagreed with the one-page statement of the ad hoc committee and that this was a very important matter. Press soon wrote back requesting that I send him the secret document and my views about it via a classified route. I complied and sent Press a classified letter of a few pages describing why I disagreed and why I thought that the ad hoc panel's conclusions were quite incorrect.

Press convened a panel of his own, in the West Wing of the White House on September 1 and 2, 1977, to explore again the evidence about yields of Soviet explosions. He called Romney to testify. The panel consisted of Springer from Livermore, seismologist Robert Massé of AFTAC, and me. Massé broke ranks with the AFTAC-ARPA view and said that Romney was incorrect about determining yields of Soviet tests. Springer, a very straightforward person, said in a nonacrimonious way that Romney was wrong. I voiced similar views. The panel concluded that the United States was overestimating the yields of Soviet explosions.

The panel recommended that the official classified magnitude-yield formula should be changed to include a correction for magnitude bias using P waves and that surface waves should be incorporated into yield determinations. It was my view that AFTAC needed to do more work on how to merge P-wave and surface-wave data in a thoughtful way. One way was to use Soviet explosions characterized by low to little tectonic release to obtain a better estimate of m_b bias. AFTAC and the Defense Department, however, did not permit that work to go forward, stating they needed to be "tasked" to do it.

I was told that Frank Press forwarded our recommendation about yield determination using both P and surface waves to Zbigniew Brzezinski, who headed the National Security Council under Carter, but that Brzezinski overruled Press and our recommendation, saying it was stupid to have two different methods for yield determination.

I also learned that the Carter administration concluded that the United States should drill deeper holes in Nevada to test weapons exceeding the 150-kiloton limit of the TTBT. This was in answer to the Soviets' alleged cheating on the treaty. I would like to think that my and a few others' raising hell and enough valid scientific points helped forestall the United States from setting off explosions above the limit of the Threshold Treaty. That would have stirred up a hornet's nest in Russia at a time of tense relations between our two countries.

MORE EVIDENCE OF BIAS IN ESTIMATING SOVIET YIELDS

By 1979 it was clear from a long and thorough publication of Marshall, Springer, and Rodean that the magnitude (m_b) bias for explosions in hard rock between the Nevada Test Site and Eastern Kazakhstan was 0.3 to 0.4 magnitude units. Their work also found that the propagation of P waves from the French test site in southern Algeria and Nevada were very similar. Romney had claimed that the Algerian site in the Hoggar (Ahaggar) region was situated in old rocks of the West African Shield and hence could be used to calibrate Soviet yields. Instead, Hoggar is one of several young elevated regions in Africa that are known as "hot spots" to those familiar with plate tectonics. It is not an old geological area like the surrounding West African Shield.

Marshall and others found that the P_n velocity beneath Hoggar itself was low, as it was beneath Nevada. Therefore, P waves were absorbed beneath Hoggar in a similar way to those beneath Nevada. Thus, Hoggar data also needed to have a correction made for a bias of about 0.3 to 0.4 magnitude units. The bottom line was that it was not a good analog for Eastern Kazakhstan.

Clearly, the British government was concerned about the U.S. methodology for determining yields of Soviet tests. Otherwise, Marshall, who worked for the UK Ministry of Defense on seismology, would not have stated his views so forthrightly in peer-reviewed journals. The same is true for Springer and Rodean—two knowledgeable scientists who had long worked at Livermore. It is quite surprising, as well as unfortunate, that their views did not carry more weight in the U.S. debate about Soviet yields. That debate festered until 1988.

A FAILED ATTEMPT AT A FULL TEST BAN TREATY

After 1963, a full test ban (a CTBT) did not resurface until the Carter administration entered into negotiations with the Soviet Union and Britain in 1977. The United States decided not to press for ratification of the TTBT and PNET but to push instead for a full ban on nuclear testing. Formal negotiations started in Geneva in July 1977. In his 1981 book, Glen Seaborg indicates that Herbert York, who led the U.S. negotiations from 1979 to 1980, informed him about what had been accomplished, most of it by the end of 1978. The main outlines of a treaty included automatic seismic detection stations on the territories of the three parties, a system of voluntary on-site inspections with arrangements for challenges and responses, a treaty with a three-year duration, and a moratorium on peaceful nuclear explosions.

After an initial good start, the CTBT negotiations stalemated when Margaret Thatcher became the UK prime minister, opposition to a CTBT increased in the United States, the Soviet Union was reluctant to accept on-site inspections, and the Soviets invaded Afghanistan. Opponents of a CTBT in the United States cited the need to test weapons for the Trident II missile and energy-directed weapons (called "Star Wars" by opponents). In addition, the Carter administration did not want to push for a CTBT until the Strategic Arms Limitation Treaty, SALT II, was completed.

10

CONTINUED DEBATE ABOUT YIELDS, ACCUSATIONS OF SOVIET CHEATING ON THE THRESHOLD TREATY, AND ITS ENTRY INTO FORCE

This chapter covers the continuing long debate in the 1980s about determining the yields of Soviet nuclear explosions. During the Reagan and first Bush administrations, the U.S. government charged that the Soviet Union had cheated on the Threshold Test Ban Treaty (TTBT) by testing nuclear weapons above its 150-kiloton limit. The issue of yield determination was finally resolved in 1988 by close-in monitoring of explosions by the United States and the USSR at the other's test sites. The United States and Russia ratified an amended TTBT, which entered into force in December 1990.

My article "The Verification of a Comprehensive Nuclear Test Ban" with Jack Evernden in the *Scientific American* of October 1982 drew criticism from DARPA officials for its conclusions about the yields of Soviet explosions, our finding that the Soviet Union had not cheated on the Threshold Treaty, and our claims that a full test ban could be monitored effectively. In the early 1980s, the Nuclear Monitoring Research Office of DARPA included Ralph Alewine, Thomas Bache, Carl Romney, and Alan Ryall, all of whom opposed a full test ban. All very conservative in their views on national security policies, they focused almost entirely on the determination of Soviet yields.

Bache wrote a DARPA report on yield determination in 1982 that stated, for yield estimation with short-period P waves, an important issue is *possible bias* caused by path effects (my italics). By path effects he meant differences in the absorption (attenuation) of P waves for paths from Soviet explosions in Eastern Kazakhstan compared to those from the Nevada Test Site. When expressed in terms of seismic magnitude, those differences are called magnitude or m_b bias, as discussed in the previous chapter.

Bias is a systematic difference, like measuring a length without realizing that an inch has been cut off the measuring stick. Just making additional P-wave measurements of magnitude (akin to measurements of length using a defective ruler) does not decrease the bias. When an average m_b is obtained from about a hundred stations, the uncertainty is very small, about 0.04 m_b units. To obtain an *accurate* estimate of the yield of a Soviet explosion from an average value of m_b, however, that magnitude still needs to be corrected for bias between it and underground explosions in Nevada. The systematic difference of about 0.3 to 0.4 m_b units between those sites dominates the uncertainty in yield determination for Soviet tests. Hence, the main task is to determine that systematic error, not to focus on more or better m_b measurements.

Bache also stated, "Regionally varying attenuation [absorption of P waves] is an important *potential* cause for bias in yield estimates [my italics]. Current knowledge suggests that differences of 0.3 or 0.4 m_b for otherwise identical events in different areas can be expected, but it is difficult to demonstrate that they actually occur."

In April 1983, Larry Burdick, a seismologist at Woodward-Clyde Consultants, wrote to me stating that Bache, his contract monitor, had reviewed Burdick's report on magnitude bias and said to him, "Your section 2 is a good example of the problems (futility?) associated with estimating a specific site amplitude bias from travel time residuals. Some important people seem to think otherwise. For example note the way Sykes and Evernden use travel time residuals to estimate m_b bias of Semipalatinsk [Eastern Kazakhstan] compared to NTS on page 55 of their article in the October *Scientific American*. I would appreciate your sending a note to Sykes drawing attention to this section in your report." Burdick then wrote to me, "Since we ultimately use travel times to estimate site bias, we obviously believe that this is a valid approach. I infer that Tom [Bache] believes otherwise." Since I had read Burdick's report, Bache was at the very least not very professional in asking Burdick to send me a note citing the errors of our ways.

In 1982 the Reagan administration decided not to submit either the TTBT or its companion Peaceful Nuclear Explosion Treaty (PNET) for ratification by the U.S. Senate until the USSR agreed to additional verification measures. Two years later, in 1984, the administration publicly charged the Soviet Union with a probable violation of the TTBT. This was part of a larger set of accusations that the USSR most likely had cheated on other arms control agreements, including the Anti-Ballistic Missile (ABM) Treaty, encryption of data during missile testing, the number of new missile systems permitted under the second Strategic Arms Limitation Treaty, and the Chemical Weapons Treaty.

Evernden and I had concluded, however, that reports of Soviet cheating on the Threshold Treaty were erroneous. When our calibration was used, it was apparent that none of the Russian weapons tests exceeded the 150-kiloton limit of the TTBT. Several Soviet explosions did come close to that limit, however, as was also the case for some U.S. tests in Nevada.

SYMPOSIUM IN 1983 ON THE VERIFICATION OF TEST BAN TREATIES

Evernden and I organized a symposium for the American Geophysical Union (AGU) on Verification of Nuclear Test Ban Treaties. This symposium, held in Baltimore in June 1983, was intended (1) to present arguments for and against verification of a complete test ban, (2) to assess the determination of yields under the Threshold Treaty, and (3) to address accusations that the Soviet Union had cheated on the TTBT.

About five hundred people attended, including many from U.S. federal agencies. A morning session was devoted to the Threshold Treaty, while the afternoon dealt with monitoring a full nuclear test ban (a CTBT). We invited Bache and Alewine of DARPA to give talks on each, which they did. Of course, Evernden and I knew to expect strong criticisms from them. I obtained a written copy of their unpublished eleven-page presentation, plus nineteen figures, titled "Monitoring a Comprehensive Test Ban Treaty."

The worst was yet to come from DARPA: an accusation that someone or some agency had leaked the purported revised U.S. classified methodology for determining yields of Soviet explosions to the media. Figure 10.1, from their presentation, showed P-wave magnitudes, m_b, of nuclear

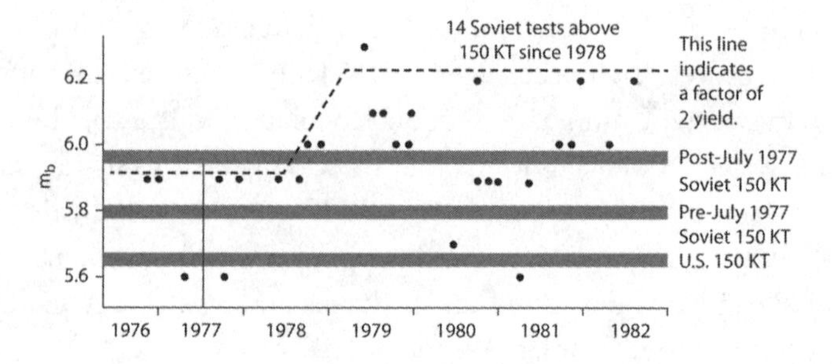

FIGURE 10.1

Seismic magnitudes m_b (filled circles) as a function of date for underground nuclear explosions at the Soviet test site in Eastern Kazakhstan.

Source: 1983 presentation by Bache and Alewine of the U.S. Department of Defense, published in Pike and Rich, 1984.

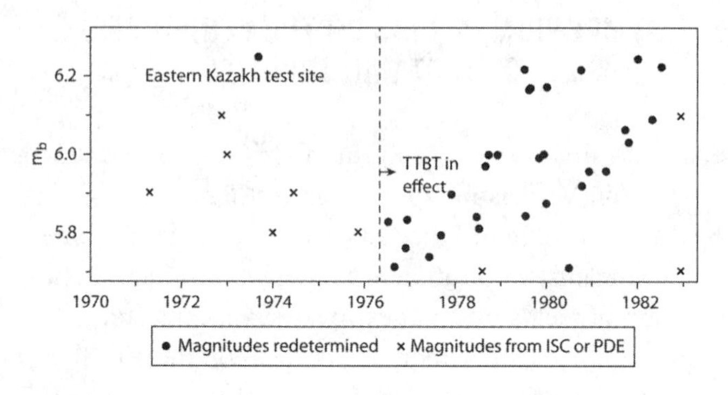

FIGURE 10.2

Seismic magnitudes m_b for underground nuclear explosions at Eastern Kazakhstan.

Source: Sykes, Evernden, and Cifuentes, 1983; also published in Pike and Rich, 1984.

explosions in Eastern Kazakhstan from the start of the TTBT on April 1, 1976, through 1982. The dates in figure 10.2, from our paper, included that period and extended it back to 1970.

The DARPA chart in figure 10.1 was annotated to show 150 kilotons for explosions in Nevada and interpretations of m_b values associated with explosions of that yield in Eastern Kazakhstan—pre and post July 1977.

It indicated that Soviet tests seemed to double in yield between 1978 and 1979 and claimed this was evidence that the Soviets had discovered U.S. plans to modify its classified yield calibration formula, presumably about July 1977. According to Alewine and Bache, the USSR raised the yields of their tests accordingly. The dotted horizontal line in figure 10.1 indicates (mistakenly) that fourteen Soviet tests after 1978 exceeded the 150-kiloton limit of the TTBT, four of them by a factor of two. Bache and colleagues published a similar figure and arguments again in 1986.

Figure 10.2, which I presented at the 1983 symposium, shows that magnitudes and yields of Soviet tests actually rose gradually over a period of three years and then stabilized at a magnitude m_b close to 6.2. Evernden, Cifuentes, and I associated m_b 6.2 with a yield of about 150 kilotons, factoring in an m_b bias of 0.35 units between hard rock in Eastern Kazakhstan and Nevada.

Figure 10.1 indicates a magnitude of 5.65 for explosions of 150 kilotons in Nevada, but this magnitude is appropriate only for tests in softer rocks, not the much harder rocks in Eastern Kazakhstan. Without taking into account the magnitude bias between the two test sites and the difference in rock types, the inference from the Department of Defense work was that Soviet tests of magnitude 6.2 corresponded to yields of about 600 to 800 kilotons. Evernden and I thought Alewine and Bache were not only incorrect but misleading as well.

After each talk at the 1983 symposium, we devoted about five minutes to questions for the speakers. Alewine received several, including one from me: "Then I have one final question that is to do with the last sentence in your abstract. And there, perhaps to paraphrase you, you liken [*sic*] yields of 600 to 800 kilotons or larger, and none of the yields you were talking about here [in your oral presentation] are in that range. Could you discuss that?" Alewine replied, "I didn't have that in the paper." I replied, "It's in your [printed] abstract." Alewine: "Right. Well, what we observed, I think what we said in the paper is that for an m_b 6.2 in the U.S. experience, we have not seen an m_b 6.2 unless yields were in the 600/800 kiloton range." Evernden responded, "But that means that the feeling that numerous people got out of the abstract, that you were concluding that possibly the Soviets had tested to 600/800 kilotons, was a false interpretation of your abstract. Is that right?" Alewine: "That's right."

In addition, I recalculated magnitudes to three significant figures (6.21, 6.14, etc.) for figure 10.2. It is clear in Department of Defense figure 10.1, however, that the DARPA scientists calculated magnitudes to only two significant figures (6.2, 6.1, etc.). Rounding off magnitudes to two significant figures is sufficient for arguing about a factor of two but not in deciding if a yield was, say, 140, 150, or 160 kilotons.

Table 10.1 lists several quotations from media coverage in 1982 and 1983 claiming that the yields of several Soviet explosions exceeded the 150-kiloton limit of the Threshold Treaty by large amounts. Some of these probably involved leaks to conservative columnists.

Why did the United States start testing near 150 kilotons soon after the TTBT became effective, but the Soviet Union did not? The United States knew from previous tests that it could detonate an explosion of 150 kilotons in Nevada without any damage occurring in the nearby cities of Las Vegas and Reno. The Soviets had conducted larger tests than magnitude

TABLE 10.1 **Published statements about yields of the largest USSR explosions under the Threshold Test Ban Treaty**

JACK ANDERSON—COLUMN IN *WASHINGTON POST*, AUGUST 1982	
350 kt	". . . the Soviets appear to have exceeded the 150-kiloton limit at least 11 times since 1978. One test in September 1980 was clocked at a likely size of 350 kilotons, according to my sources."
260 kt	"As recently as July 4, the Soviets set off a huge nuclear blast. It was estimated at a likely 260 kilotons, . . ."
HAROLD M. AGNEW—LETTER TO *SCIENCE*, APRIL 8, 1983	
400 kt	". . . subsequent tests appeared to us to range as high as 400 kilotons . . ."
ALEWINE AND BACHE—*EOS*, MAY 1983 (ABSTRACT FOR SYMPOSIUM ON JUNE 2)	
600–800 kt and larger	"In U.S. experience an m_b greater then 6.2 (as measured for the largest Soviet events) has only been seen for yields of 600–800 kt and larger."
JUDITH MILLER—*NEW YORK TIMES*, JULY 26, 1982	
300 kt	"One official said that there had been several Soviet tests, many at one particular site, that had been estimated at 300 kilotons."

Source: Sykes and colleagues, 1983.

6.2 before the TTBT became effective in 1976 at their remote Arctic test site at Novaya Zemlya. If they had been set off in Eastern Kazakhstan, their stronger shaking likely would have caused damage in the nearby city of Semipalatinsk. This may explain the gradual increase in yields over three years (figure 10.2) to ascertain how much damage larger tests would cause in that city.

Soviet explosions larger than about magnitude 6.2 or 150 kilotons might well have produced damage in Semipalatinsk. In 1984 Evernden and I claimed that if a 75-kiloton explosion were set off at the site of the 1964 *Salmon* explosion in Mississippi, it would have caused damage to nearby towns and cities. Our reasoning was that propagation of seismic waves would be as damaging in Kazakhstan as it would be in Mississippi (or the rest of the central and eastern United States). A recent example of efficient wave propagation and strong shaking can be seen in the damage to the Washington Monument and the National Cathedral in Washington, DC, which was 100 miles (160 km) away from the magnitude 5.8 earthquake near Mineral, Virginia, in August 2011. Soviet explosions in Eastern Kazakhstan of 75 kilotons generated seismic waves of about m_b 6.

At the 1983 symposium, Inés Cifuentes, a graduate student at Lamont, and I presented new work on the estimation of Soviet yields at the Eastern Kazakhstan test site from 1978 through 1982 using digital recordings of seismic surface waves. Our work, published in 1984, was prompted by a letter I received on December 15, 1982, from Peter Marshall, who had long worked on seismic verification in Britain, about my use of surface waves for yield determination in our 1982 *Scientific American* paper. Marshall wrote, "I have always thought that [the surface wave magnitude] Ms should be a stable indicator of yield. Some LR [long-period Rayleigh wave] trains from the RTS [Russian Test Site in Eastern Kazakhstan] have shaken my confidence in that there are examples of where the average Ms has been significantly reduced in amplitude in all azimuths [directions]."

What concerned Marshall and others, understandably, was that some explosions at that site triggered the release of large amounts of natural stress in the surrounding rocks. The tectonic release at the time of explosions in hard rock, depending on the amounts of natural compressive stress built up before hand, reduces the size of the surface waves from shots in Eastern Kazakhstan. Because those surface waves are reduced in amplitude, the yields computed from them from Ms are too small. Tectonic stresses in Nevada, however, are extensional, not compressive, and have the opposite effect on the size of surface waves.

Cifuentes and I dealt with tectonic release in a study of twenty underground explosions at Eastern Kazakhstan. We used the ratio of the amplitudes of two surface waves—Love waves/Raleigh waves—as a measure of the amount of tectonic release that was contaminating Raleigh waves produced by the explosions themselves. Love waves are generated only by natural tectonic stress release, not the explosion itself. We calibrated a new m_b-yield relationship for Eastern Kazakhstan by calculating yields from surface waves, but only for those explosions that were characterized by a small-to-negligible component of tectonic stress. We then used that relationship to determine yields from m_b for all twenty explosions.

We concluded, "The yields of the seven largest Soviet explosions are nearly identical and are close to 150 kilotons, the limit set by the Threshold Treaty." For explosions characterized by a large Love to Raleigh wave ratio—that is, a large tectonic component—we recommended using just our new m_b-yield relationship to determine their yields from body waves (m_b) alone. This is what I had hoped and expected that AFTAC would do in 1977, but they did not.

Our position was slowly becoming clear to journalists. John Wilke wrote about the AGU symposium in the *Washington Post* on June 3, 1983. He mentioned a classified study at the Livermore Lab, stating that a Livermore physicist said publicly to him, "Whatever it is believed in Washington, it is now clear that officials here at Livermore Lab do not believe that the Soviets have violated the 150-kiloton limit."

Journalist R. Jeffrey Smith, who also covered the symposium, quoted seismologist Bernard Minster of UC San Diego in the June 17, 1983, issue of *Science*: "Based on what I heard this morning, I think we have a hard time justifying statements that the Soviets are cheating." Robert North, another consultant to the Defense Department, said, "After listening to the presentations . . . most people would agree that you cannot assert that the Soviets have violated the Treaty." Smith went on to state, "Milo Nordyke, who directs the verification program at Lawrence Livermore National Laboratory, where the bulk of the government's analysis is conducted, said after the AGU meeting 'DARPA, which takes the most conservative view, certainly seems to be in the minority. Of course, with the conservative view, you automatically get some evidence of Soviet violations. But you have to use the best estimate, not the most conservative one. This is a message that the politicians in Washington have a hard time understanding.'" Pike and Rich of the Federation of American Scientists also quoted Nordyke as saying "there is no hard evidence of Soviet test violations."

DARPA and AFTAC supported several analyses of the attenuation of short-period P waves beneath test sites in Eastern Kazakhstan, Nevada, Mississippi, Amchitka Island in the Aleutians, and Algeria. In January 1984 Robert Blandford and colleagues at the consulting firm Teledyne Geotech computed corrections to m_b magnitude for each site with respect to Nevada. They reported that an m_b of 6.17 was associated with 150-kiloton explosions at Eastern Kazakhstan, very similar to the results we presented in 1983. They found that P waves from French explosions at the Algerian hot spot were attenuated by about the same amount as those in Nevada and the attenuation of P waves beneath Amchitka was similar to those beneath Eastern Kazakhstan.

DARPA PANEL MEETINGS ON YIELD DETERMINATION

In January 1983, I was invited to be a member of a Department of Defense Technical Review Panel on Threshold Test Ban Treaty Verification Issues that was convened by DARPA. After attending its first meeting, I informed DARPA that I would be out of the country and not able to attend the next meeting set for about July 29. I stated, "I have had a fairly long telephone conversation with Gene Herrin, and will send him written comments in response to his unclassified letter. I also think that it is important that my views be represented along with those of Sean Solomon [of MIT]. I understood that Sean also will not be able to attend."

On August 17, 1983, Alewine, who had become director of the Geophysical Sciences Division of DARPA, sent me a letter informing me that a separate classified package of the completed panel report had been mailed to me. His unclassified letter states, "As we discussed in the Panel meetings, the methodology for the use of surface waves in yield estimation has not received the level of critical review as has that for body waves."

He wrote, "We would like to convene the Panel about mid-October to examine fully the status of using surface waves for TTBT monitoring . . . we would like for the Panel to begin review of technical aspects of Comprehensive Test Ban Issues at the October Panel meeting."

I wrote an unclassified letter to Herrin on September 16, 1983, with my general views about the Panel Report and a classified letter to DARPA about some specific points. I stated, "I conclude that the panel's recommended value for the bias [in magnitude between the Nevada and Eastern Kazakhstan] is still too low. A consequence of this, of course, is that calculated yields will still be too high." On May 31, 1983, Under Secretary of Defense Richard D. DeLaure wrote to Frank Press, then the president of the National Academy of Sciences, stating that he foresaw no need for a parallel review effort on questions of verifying nuclear test bans to be conducted by DARPA and that all legitimate questions of objectivity and credibility were well met by the [present] DARPA panel."

I also stated in my letter, "I am seriously concerned about the procedures that have been followed in previous panels of which I was a member. Let me be specific on three points. DARPA officials, Bache, Alewine, were present at all of the sessions of the panel at its January 1983 meeting. They have consistently and often reiterated their views that the bias between those two test sites is small. They have actively and aggressively participated in the deliberations of the panel as if they were members. Under those circumstances, I believe that it is difficult for different views to be heard and to be considered in a thoughtful manner.

"Secondly, the panel has adopted positions that are at odds with those of several distinguished scientists including Dr. Peter Marshall, of the United Kingdom, Springer, Rodean and [Peter] Moulthrop of Livermore, Evernden, etc. I believe that none of them has had a chance to respond to criticisms of their work made in either the latest or earlier reports of the panel. Thirdly, . . . my sense is that the panel has had too little time during its meetings to write a thoughtful and independent report of its own."

I sent copies of my letter to Alewine and the panel members. I was not invited to the October 1983 meeting. I asked panel member Tom Jordan of the University of Southern California if I was still on the panel, and he said no. I never received a letter from DARPA to that effect. I was not invited to be a member of future DARPA or AFTAC panel meetings.

The DARPA panel continued to meet until at least 1985, because I received an unclassified draft summary of their work dated February 1985. It discussed the contamination of surface waves from tests in Eastern Kazakhstan and recommended not using them for yield determination. They did not acknowledge the significant problems of using just P waves for yield determination. I was told that members of the panel disagreed about magnitude bias and wanted sections put in about it, but that material was removed from the final report.

Writing in the *New York Times* on April 2, 1986, Michael Gordon, the military affairs correspondent, stated, "The Central Intelligence Agency has changed its procedures for estimating the yields of large Soviet nuclear tests because it has decided its previous estimates were too high, Reagan Administration officials said today. . . . Experts familiar with the change said it would lower estimates of the yield of Soviet tests by about 20 percent."

Gordon went on to say, "On Oct. 18, a panel of scientists selected by the Defense Advanced Research Projects Agency prepared a classified report that concluded the Government's method for estimating the yield of Soviet explosions was based on faulty assumptions. The panel's report was submitted in late October to the Joint Atomic Energy Intelligence Committee, which issues reports on the size of foreign nuclear explosions. The committee is made up of members from the military services and intelligence agencies. . . . On Dec. 17, the Joint Atomic Energy Intelligence Committee recommended that the C.I.A. adopt the advice in the report commissioned by the research agency. Officials said the Defense Intelligence Agency disagreed, but was overruled.

"Officials said applying the new method retroactively would still leave about a dozen Soviet tests that appear to be above the limit, and one official said only three or four of these exceeded the limit enough to warrant special concern." These statements imply that the bias in magnitudes for Eastern Kazakhstan was about 0.2, not 0.3 to 0.4 as several of us had recommended. I had assumed, apparently incorrectly, that the U.S. government had corrected the formula for estimating Soviet yields in 1977.

Gordon's article indicated that it did not occur until 1986. U.S. formulas for estimating Soviet yields have changed with time and are still classified.

Gordon also stated, "Richard N. Perle, the Assistant Secretary of Defense for International Security Policy, reportedly opposed adopting the recommendations and argued that the issue needed more study. Mr. Perle declined to discuss the issue. Administration experts, who asked not to be identified, were divided about whether the change should lead the Administration to drop its allegations against the Soviet Union."

PERLE AND THE SCIENTISTS

Brian McTigue produced a television interview in 1986 for San Francisco station KRON called "Richard Perle and the Scientists: The Controversy Over Nuclear Testing." They filmed it during the debate about alleged Soviet cheating on the Threshold Treaty. It focused on comments made by seismologists Charles Archambeau of the University of Colorado, James Hannon of Livermore, and me, countering statements by Perle. Perle served in the Reagan administration from 1981 to 1987 as assistant secretary of defense for international security policy and later as chairman of the Defense Policy Board, an influential group of advisers to the Pentagon. The program can be viewed on youtube.com.

I single out Perle here because he was a hard-liner and the strongest opponent of the control of nuclear weapons for more than fifty years. He exaggerated Soviet nuclear capabilities on many occasions. What is particularly relevant here is that he knew nothing about seismology, which was at the very heart of estimating Soviet nuclear yields and possible cheating. Congressional, arms control, and scientific communities knew him as the "Prince of Darkness."

Here are excerpts from the 1986 interview.

INTRODUCTION: The Soviet Union has stopped testing and is abiding by a nuclear test ban, but the United States has continued to test weapons underground claiming the Soviets have cheated. The Administration says the Soviets tested weapons more powerful than agreed to in 1974, but have they?

Rollin Post and our Target 4 Investigative Unit found the Administration is ignoring evidence that the Soviets never cheated, that they are following the treaty.

CHARLES ARCHAMBEAU [SEISMOLOGIST, COLORADO]: If the scientific data doesn't quite agree with your political position, what is done is to bend the data a little bit.

PERLE: Baloney! It is not a question of scientific evidence. It is a question of scientists playing politics. I've looked carefully at the evidence and have concluded as President Reagan did that there is significant evidence that the Soviets have violated the 150-kiloton threshold.

ROLLIN POST [INTERVIEWER]: This has been the position of the Administration since 1983. But it has also caused a rebellion among the very scientists the Defense Department relies on to estimate the size of the Soviet tests. Target 4 interviewed some of those seismologists and they all said the Soviet Union has not violated the 1974 Test Ban Treaty.

ARCHAMBEAU: At present there is no evidence that the Soviets have tested over 150 kilotons, none whatsoever.

LYNN SYKES [SEISMOLOGIST, NEW YORK]: The treaty itself states that neither country should test above 150 kilotons and I have no evidence that indicates to me that the Soviets have done that.

WILLARD HANNON [SEISMOLOGIST, LIVERMORE NATIONAL LABORATORY]: I don't believe that the evidence supports a militarily significant violation.

PERLE: The best experts available spent years studying this and came to the conclusion that it was likely the Soviets had violated the 150-kiloton threshold.

Well, with all due respect he [Archambeau] is wrong, there is lots of evidence. He may not be persuaded on the basis of the evidence, but to say that there is no evidence is just flatly wrong.

POST: Lynn Sykes of Columbia says that the Soviet Union has not violated the 150-kiloton limit of the threshold treaty as alleged. He's wrong?

PERLE: He's entitled to his opinion. He's a professor sitting up at Columbia.

Well, all that seismology enables you is to make an estimate as to the yield of an event. Even by that standard alone, there is evidence that suggests the Soviets have violated it, your experts not withstanding. There is other evidence as well that is sensitive and of a classified nature.

POST: Mr. Perle would only say the other evidence involved satellite and electronic surveillance. Target 4's investigation learned that Mr. Perle had already convened a panel of experts, which looked at this other evidence and rejected it.

PERLE: They came to the conclusion that of the many ways of estimating yield, seismology was the single most important and I happen to agree with that.

I did not particularly care much what their answer was, it did not have any profound bearing on our policy.

POST: Target 4's investigation has also uncovered evidence that Mr. Perle improperly tried to manipulate intelligence agencies in a biased direction. Example: Perle's letter to the Air Force when its intelligence unit asks seismologists to advise on Soviet nuclear tests. According to sources who have seen the letter, it said the intelligence community is undermining the Administration's position. My Department will control this area. I asked Perle about the letter.

PERLE: I don't remember the exact words of the letter, but my concern was the concern that I have been expressing to you throughout this interview, which is that we have tended, I think wrongly, to exclude the non-seismic evidence that bears on the estimation of the yield of Soviet tests.

They're all seismologists; they're a bunch of seismologists feathering their own nests. Well, seismologists have dominated this field from the beginning. It's how they make their living. The day that it is concluded that we can get along without attributing the importance to seismology that we do—some of these fellows are going to be looking for jobs.

ARCHAMBEAU: The scientific opinion is close I'd say to unanimous. Right now Mr. Perle finds it extremely difficult to find any scientists that will defend the DoD Perle position and that's because there just aren't any that believe it.

SYKES: I think that one view that is often put forth is that arms control agreements are not in the best interest of the United States. That the Soviets will cheat and then attempt to have a self-fulfilling prophecy by coming up with some procedure, an incorrect one, that indicates to them that the Soviets have cheated.

In one of Perle's statements, he commented that he had additional classified information that led him to different conclusions about verifying

Soviet nuclear testing. His responses to similar previous statements had led people with high-level clearances for all of the relevant documents to reexamine them. Perle was found wanting.

In a later frightening interview on the Public Broadcasting Service's program *Frontline*, Perle made the case for using a war with Iraq to remake the Middle East. He stressed the significance of the World Trade Center disaster of September 11, 2001, in shaping the Bush administration's thinking about the links between terrorism and weapons of mass destruction.

SENATE AND HOUSE ACTIONS ON NUCLEAR TESTING

The U.S. House of Representatives and the Senate, which were controlled by Democrats, held hearings on Soviet compliance with the Threshold Treaty and verification of a full test ban in 1985 and 1986. I gave oral and written testimony on both topics, twice to House subcommittees and once to the Senate Foreign Relations Committee.

Alewine sent a letter on December 16, 1985, to Representative Beverly Byron of the House Committee on Armed Services criticizing several statements in my testimony before her subcommittee on November 20, 1985. Alewine said, "Dr. Sykes represents one extreme in the assumptions he advocates (leading to lower yield estimates than most). There are others, just as responsible and knowledgeable, who advocate extreme assumptions on the opposite end (leading to higher yield estimates). The reviews conducted by the DoD have tended to balance these extremes." He made no mention of the several scientists and officials at Livermore and Britain who disagreed with his views. Alewine also stated, "The flippant remarks of Dr. Sykes concerning the possibilities for cavity decoupling evasion gloss over a very difficult and threatening problem that must be addressed."

In the 1980s I debated Alewine several times about Soviet yields and the verifiability of a full test ban treaty. I had no doubt that Alewine, Bache, Perle, and others consistently played "hard ball." Someone said that if you want to argue with those people, you have to be willing to jump into the pigpen. The debate went on for a very long time, and I am glad I kept at it. My testimony in 1985 and 1986 likely contributed to Congress's taking an active role in the verification of a full test ban and ascertaining if the Soviet Union had cheated on the Threshold Treaty.

INDEPENDENT REVIEW BY THE OFFICE OF TECHNOLOGY ASSESSMENT

The Senate Select Committee on Intelligence, the House Committee on Foreign Affairs, and the House Permanent Select Committee on Intelligence requested that Congress's Office of Technology Assessment (OTA) undertake an assessment of these test ban issues in 1986. OTA had previously performed studies for Congress on arms control and other scientific and technical issues.

OTA conducted the first major independent review on seismic verification of nuclear testing from 1986 to 1988. For the first time, the departments of Defense and Energy were *not* the sole sources of information in the government.

OTA formed an advisory panel of nineteen individuals from various government agencies, universities, the weapons labs, and consulting firms. Paul Richards of Lamont, who had been involved in test ban issues for some time, and I were members. Gregory Van der Vink of OTA was the project director. He received a PhD in geophysics from Princeton and became interested in science policy, especially arms control, through Frank von Hippel and his group at Princeton. Frank had long worked on national security issues and energy. Greg persuaded OTA to conduct the test ban study.

The study itself and its 1988 publication *Seismic Verification of Nuclear Testing Treaties* were approved by the Technology Assessment Board of Congress, which consisted of twelve members chosen equally from Republicans and Democrats and from the House and Senate.

The 1988 report stated, "Seismic monitoring is central to considerations of verification, test ban treaties, and national security." It addressed two key questions:

1. Down to what size explosion could underground testing be monitored seismically with high confidence?

2. How accurately could the yields of underground explosions be measured seismically?

The answers to these questions would provide the technical information that lay at the heart of the political debate over:

1. how low a threshold test ban treaty with the Soviet Union we could verify;
2. whether the 1976 Threshold Test Ban Treaty was verifiable; and
3. whether the Soviet Union had complied with present testing restrictions.

OTA held workshops of two days each on (1) Seismic Network Capabilities, (2) Identification, (3) Evasion, and (4) Yield Determination. About a dozen individuals participated on each panel. Richards and I were on the panels on Identification and Yield Determination; Richards was also on Network Capabilities, and I was also on Evasion. While some of the briefings were at the secret level, the 1988 report was cleared by relevant government agencies prior to its open publication. The report is a good primer on seismology as it relates to nuclear monitoring and on the role of verification in the context of national security.

In its executive summary, the report concluded that unless differences in the transmission of seismic P waves beneath Eastern Kazakhstan and Nevada were taken into account, the sizes of Soviet explosions were greatly overestimated. Once these differences were appropriately considered, the report stated, "All of the estimates of Soviet and U.S. tests [since the TTBT became effective in 1976] are within the 90 percent confidence level that one would expect if the yields were 150 kt or less. Extensive statistical studies have examined the distribution of estimated yields of explosions at Soviet test sites. **These studies have concluded that the Soviets are observing a yield limit consistent with compliance with the 150 kt limit of the Threshold Test Ban Treaty** [original boldface]." I consider this an important victory for good science, sound science policy, and arms control in general.

The 1988 report has a table on page 124 showing my calculations of the yields of the six largest explosions at Eastern Kazakhstan since 1976. All of those yields were close to the 150-kiloton limit and well within the uncertainty expected for its observance.

ON-SITE MEASUREMENTS OF YIELD

Two non-seismic methods are available for determining the yield of explosions near the 150-kiloton limit of the Threshold Treaty. One involves drilling back into the explosion point and obtaining samples of various radioactive materials produced by the explosion. This method, often called radiochemical, or rad-chem, has an uncertainty of about 10 percent. Although the United States often used rad-chem for its tests, applying them to explosions on-site could reveal information about the characteristics of weapons that the Soviets and Americans likely wanted to keep secret.

Another method for yield determination, abbreviated CORRTEX, involves drilling a second hole very close to and equal in depth to that used for a large underground explosion. When the nuclear explosion occurs, a cable in the second hole is crushed. This method measures the speed of the shock wave, which travels faster than the speed of sound, close to a nuclear explosion as it crushes the cable. The satellite hole must be within about 33 feet (10 meters) of the hole containing an explosion with a yield of about 150 kilotons.

These measurements permit yield to be determined with an uncertainty of about 30 percent. This method, of course, is very intrusive because it is so close to the explosion. Its advantage is that it does not reveal the radiochemical contents of the materials produced by the explosion. It is not applicable, however, to small nuclear explosions. The 1988 report contains an extensive appendix on this method. An idea was for the United States to use CORRTEX to monitor a Soviet test and for them to use similar equipment for a U.S. test in Nevada.

By 1986 General Secretary Gorbachev of the Soviet Union had put President Reagan under pressure to push for a full test ban. Reagan proposed that if Gorbachev would agree to work on the verification of the TTBT and PNET first, the United States would then negotiate ways to implement a step-by-step program limiting and ultimately ending nuclear testing. Reagan said, however, that he would not ratify the TTBT even if the Senate gave its advice and consent.

During expert talks in July 1987, the Soviets proposed calibration of test sites to reduce uncertainty in yield estimation. They invited the

United States to measure yields at one of their main sites using CORRTEX and regional seismic observations. The two countries then signed a bilateral agreement to conduct Joint Verification Experiments (JVEs) in 1988 at Eastern Kazakhstan and Nevada. Each country was to detonate one explosion with a yield between 115 and 150 kilotons.

The United States did not release the yield of the Soviet JVE test of September 14, 1988, which it monitored using CORRTEX. Each country released to the other the yields of the JVE and five previous tests at each site. The 1988 agreement stated that yields were not to be released to additional people or countries without the consent of the other. As far as I know, these yields were kept secret, and still are. Nevertheless, a detailed report on the two Joint Verification Experiments in the *New York Times* by Michael Gordon states that American and Soviet on-site measurements were said to give yields of 115 and 122 kilotons, respectively, for the Soviet JVE, an average of 118.5 kt.

In 1989 Göran Ekström, a geophysicist then at Harvard, and I published an average magnitude, m_b, of 6.115 with a very small uncertainty of +/– 0.018 for the Soviet JVE using reports of P waves from sixty-eight seismic stations. We then used three m_b-yield relationships derived from various test sites where yields were available. Each was corrected for m_b bias with respect to Eastern Kazakhstan. We chose a bias of 0.35 for explosions in hard rock and those below the water table in Nevada. The three calibrations gave yields very similar to those reported in the *New York Times* for the Soviet JVE. We extrapolated the yield of 118.5 kilotons measured on-site for the Soviet JVE to 150 kilotons. It gave a magnitude m_b of 6.20, very similar to the magnitudes of the six largest explosions at that test site published in the 1988 OTA report.

YIELDS OF SOVIET NUCLEAR TESTS MEASURED USING LG SEISMIC WAVES

A third method of determining yields from seismic waves came of age in 1992. In 1973 Otto Nuttli of St. Louis University started to develop a magnitude scale called m_{bLg} that uses short-period seismic waves called Lg with periods near one second (one cycle per second frequency).

In 1952 Press and Ewing of Lamont had described and named Lg and another slow surface wave called Rg, which propagate in continental areas. Lg is often the largest wave on a short-period seismic record from earthquakes to stations at regional distances within continents. P waves from explosions, however, are relatively large compared to Lg, making Lg a good method for identifying underground explosions and potentially for determining their yield.

Nuttli developed m_{bLg} magnitudes suitable for all parts of the United States. In regions of older crust, Lg is observed at distances of more than 3000 miles (5000 km). Examples of this are Lg recordings at Palisades, New York, of waves that cross the ancient rocks of Canada from earthquakes in the Yukon and northern Alaska.

Lg radiates symmetrically from explosions and earthquakes, and it does not have the major difficulty of P waves, which need a correction for differences in wave propagation through the uppermost mantle of the Earth. Hence, it became very useful for yield determinations.

In 1992 Frode Ringdal of Norway, Marshall of Britain, and Alewine of DARPA obtained more precise estimates of m_{bLg} than Nuttli had for explosions in Eastern Kazakhstan using data from the NORSAR (Norway) and Grafenberg (Germany) seismic arrays. Instead of measuring the amplitude of a single wiggle of Lg, they obtained an average over several minutes. Lg, as shown in figure 10.3, is not a single pulse but a train of seismic waves that builds up and then decays slowly. Its records are more like those of waves that travel great distances in the oceans, called the T phase, and those made on the moon.

Ringdal and colleagues analyzed seismic data for 101 nuclear explosions at the Shagan River portion of the Eastern Kazakhstan test site, which was the site of most large underground tests. Their measurements of m_{bLg} and m_b determined from P waves differed systematically by as much as 0.15 magnitude units between two subareas of the Shagan River site. They discovered that m_b values for four explosions with yields published in 1989 by V. S. Bocharov and his Russian colleagues also varied in the same way with respect to values of m_{bLg}. Those subareas are separated by a series of major faults oriented northwesterly. Hence, Ringdal and his colleagues concluded that m_{bLg} is a more reliable measure of yield than either m_b or surface waves.

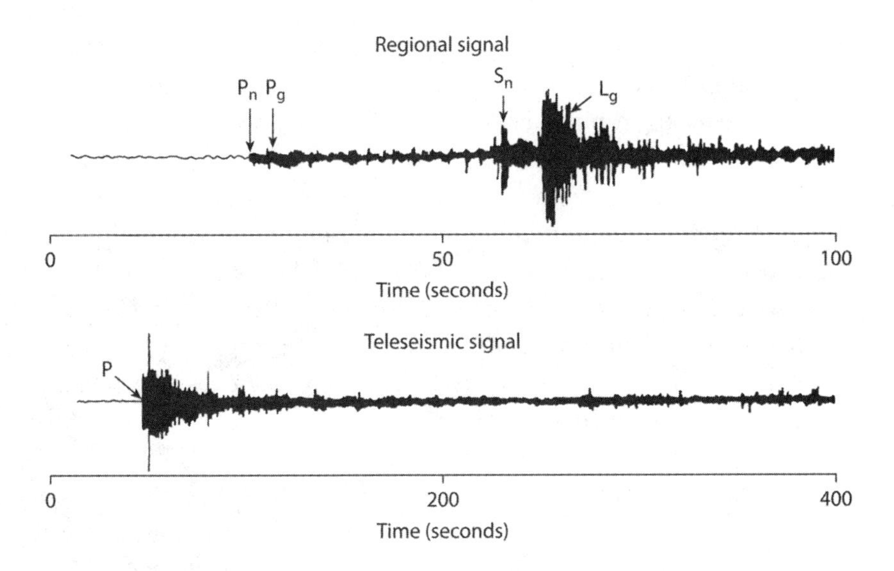

FIGURE 10.3

Upper seismogram is from an event at a regional seismic station that propagated along a continental path. L_g is its largest signal. Lower seismogram is from an event to a station at a large distance (a teleseismic signal). The P wave is its largest signal, and L_g is very small.

Source: Office of Technology Assessment, 1988.

They found a magnitude m_b bias with respect to Nevada of 0.45 for the southwestern part of that area and 0.30 for the northeastern part. These are close to the biases of 0.3 to 0.4 that several of us determined earlier, and clearly not zero. By subdividing that test site into two parts, Ringdal and colleagues provided better determinations of yields than any of us had obtained previously. They obtained an average yield for the Soviet JVE explosion of 1988 in the southwest part of the test site of 108 kilotons, which compares very well with the 118.5 kilotons reported in the *New York Times* for the two close-in CORRTEX measurements.

A lingering question is whether the Soviet Union has cheated on the Threshold Treaty by conducting any tests above its 150-kiloton limit since its start date in 1976. It is important to understand that all yield determinations have uncertainties associated with them. For explosions of 150 kilotons, some estimates will be somewhat larger, others somewhat smaller, and some right on, as shown for the three following explosions with yields near 150 kilotons.

Only one of the explosions studied by Ekström and Richards, that on April 3, 1987, for which they calculated a yield of 176 kilotons, exceeded the 150-kiloton threshold. Ringdal and colleagues, however, obtained 140 kilotons for that explosion using m_b and m_{bLg}. For a second explosion, on August 4, 1979, the two sets of authors obtained 153 and 132 kilotons. For a third event, on October 27, 1984, they calculated 165, 104, and 140 kilotons. Those measurements taken together indicate that the three largest explosions at Eastern Kazakhstan testing area since 1976 were at or close to the 150-kiloton limit within the uncertainties of the measurements.

When asked by a member of a subcommittee of the U.S. House of Representatives in 1985, I stated that the yields of Soviet tests in Eastern Kazakhstan near the threshold of the treaty could be determined [then] with an uncertainty of about 30 percent. Donald Kerr, the director of Los Alamos, said my estimate of the uncertainty in yield of 30 percent was small compared to extrapolations that could be made up to a factor of about two times. With the introduction of Lg measurements, the uncertainty in yield estimation for Eastern Kazakhstan was reduced further to about 25 percent.

To appreciate what these numbers mean in terms of accuracy, a person driving 25 percent faster than a 55 mile per hour speed limit would be traveling at 69 miles per hour and might or might not receive a speeding ticket. If the person were traveling 25 percent slower, the speed would be 44 miles per hour, and he or she certainly would not get a ticket. If Russia had decided to test weapons at yields 25 percent higher than the 150-kiloton limit, or 187 kilotons, the yield determined by the United States likely would be between 150 and 234 kilotons but more likely close to 187 kilotons. The analogy is that they probably would be "given a speeding ticket"—that is, accused of cheating on the TTBT. More than one test at 187 kilotons would have substantially increased the chances of the United States' determining that the Soviet Union was testing above the threshold of the TTBT.

Alewine joined Ringdal and Marshall as third author of their 1992 paper. This probably was an unstated acknowledgment by Alewine, as well as by the U.S. Defense Department, that the determination of yields at that test site had been resolved. It was clear that earlier U.S. charges of Soviet cheating, spearheaded by Alewine, Bache, Perle, and Romney, were false. The Soviets, in fact, had been in compliance with the 150-kiloton limit of the Threshold Test Ban Treaty. With that, the "yield wars" were finally over. Alewine told me in 2009, at a meeting in Vienna on nuclear test testing, "I guess you got about 50 percent of things right and we [DARPA] 50 percent." I would give them no more than a 5 percent.

In 1989 Bocharov, Zelentsov, and Mikhailov of the Soviet Union published an official list of ninety-six underground nuclear explosions at the Eastern Kazakhstan test site through 1972, before the Threshold Treaty became effective. Mikhailov became minister of atomic energy of the Russian Republic in 1992. While most yields are listed within a broad range, several are given exactly. They list two larger than 125 kilotons: 165 kilotons on November 2, 1972, and 140 kilotons on December 10, 1972. My determinations of those yields were 154 and 138 kilotons; Ringdal and colleagues calculated 169 and 158 kilotons. I estimated the yield of the largest underground explosion at that test site, on July 23, 1973, as 193 kilotons. All three of those explosions, of course, occurred before the Threshold Treaty became effective in 1976.

From 1990 to 1992, physicist David Hafemeister worked for the Senate Foreign Relations Committee to examine arms control treaties at the end of the Cold War. In an article published in 2005, he stated, "This charge [of a probable violation of the Threshold Test Ban Treaty] was removed in 1990 after the 1988 CORRTEX measurements at Semipalitinsk [Eastern Kazakhstan] Test Site and after properly taking into account the geological differences between test sites. . . . The U.S. record on TTBT noncompliance charges was not entirely honorable."

By 1990 the United States and Russia completed a considerably revised protocol for the Threshold Test Ban Treaty. I doubt if more than a few people ever read the new protocol, which is exceedingly long. It called for CORRTEX-type measurements for U.S. and Russian weapons tests larger than 50 kilotons. Because the Russians stopped testing before the treaty entered into force in late 1990, the United States has not been able to make additional CORRTEX measurements of Russian nuclear explosions. Under the ratified TTBT, the Russians, however, were able to monitor two U.S. explosions in Nevada before President Clinton halted U.S. testing in 1992.

The TTBT and its companion Peaceful Nuclear Explosions Treaty (PNET) are still in force.

A SOVIET ASSESSMENT OF THE THRESHOLD TEST BAN TREATY

Roland Timerbaev of the Russian Ministry of Foreign Affairs, who attended the negotiations for the Threshold Treaty in 1974, wrote in 2006 about the history of the treaty. In a long footnote, he states that the yield of the U.S. Joint Verification Experiment (JVE) of 1988 in Nevada was 180 kilotons and significantly exceeded the 150-kiloton limit of the treaty. He quotes an interview on December 7, 2005, in which V. N. Mikhailov of the Russian Ministry of Atomic Energy stated that a collapse depression was formed by that JVE. Timerbaev states, "I might add that the Americans were very upset about this and requested that we not speak about the matter publicly; however, the story has since become public knowledge through the media." Springer and others list the yield of the U.S. JVE, called *Kearsarge*, as 100–150 kilotons and do not mention a collapse feature.

Timerbaev also gives information about how the yield threshold for the TTBT was debated in 1974, which I did not know. He states that the Soviet Ministry of Defense and the agency responsible for building and testing Soviet weapons (called at the time the Ministry of Medium Machine Building or Minsredmash) pushed to allow the USSR to undertake one or two tests [presumably per year] with yields over a megaton and three to four of 500 kilotons.

Timerbaev states that it seemed to him during the [1974] negotiations that the threshold would be set in the range of a few hundred kilotons. He quotes a conversation in 1984 with a former American official reporting that in 1974 the U.S. military wanted to establish a threshold at 600 kilotons but Henry Kissinger, Nixon's secretary of state, sharply objected. Kissinger and Gromyko agreed upon a 200-kiloton threshold, but Nixon, at Kissinger's behest at the last minute, wanted a 150-kiloton limit. That number was agreed upon in the last few days of the negotiations in Moscow just before Nixon and Brezhnev completed and signed the treaty in early July 1974.

The *Mike* thermonuclear (hydrogen bomb) explosion. Photo by U.S. Atomic Energy Commission.

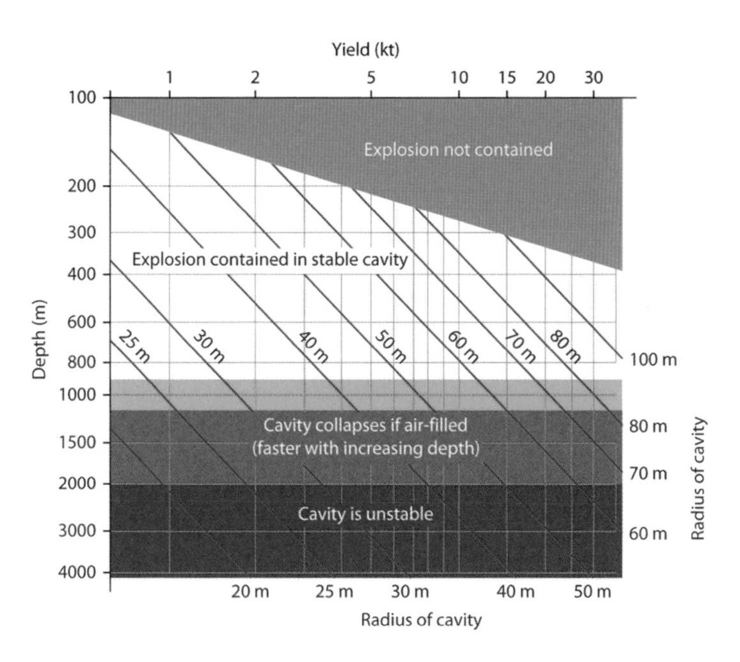

Depths (vertical axis) for conducting decoupled nuclear explosions in an air-filled cavity in salt. Yield of the explosion is on the horizontal axis at top; the cavity radius is at bottom. Modified from Davis and Sykes, 1999.

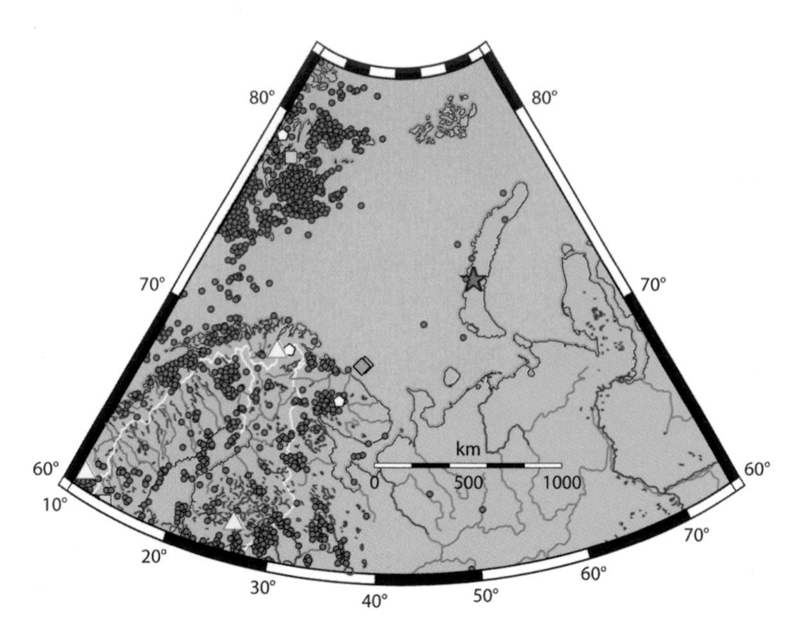

Seismic events and stations in the vicinity of the main Russian arctic test site at Novaya Zemlya. Red stars show the locations of nuclear tests since 1977. Small blue circles indicate seismic events from 1999 to 2009 of magnitude greater than 2.0. The primary stations of the International Monitoring Service (triangles), the auxiliary station in Sweden (square), and three other publicly available stations (pentagons) are shown in white. The orange diamond locates the Kursk submarine disaster of 2000. Many of the events in Scandinavia, Finland, and mainland Russia were small mine blasts. *Source*: National Academies Report, 2012.

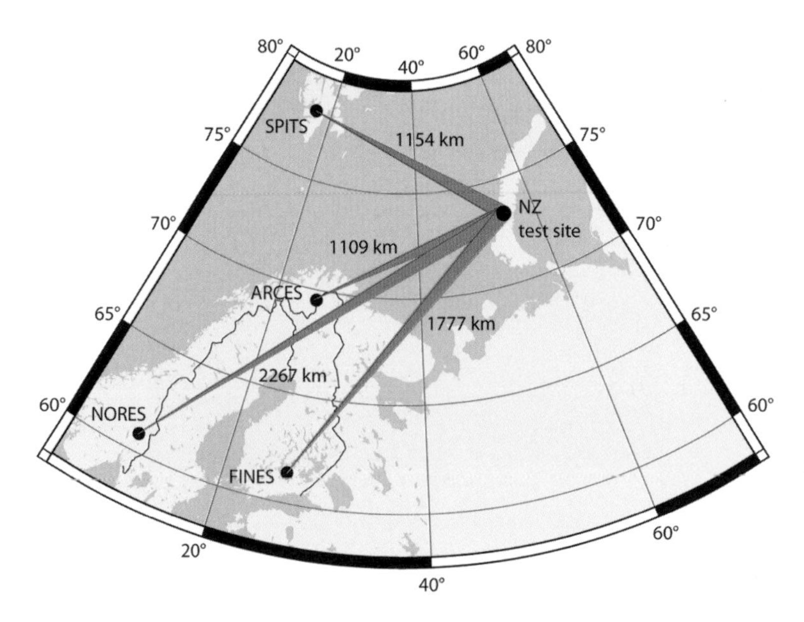

Locations of four seismic arrays that continuously monitor the Russian test site on Novaya Zemlya (NZ). *Source*: Kværna, unpublished figure 2016.

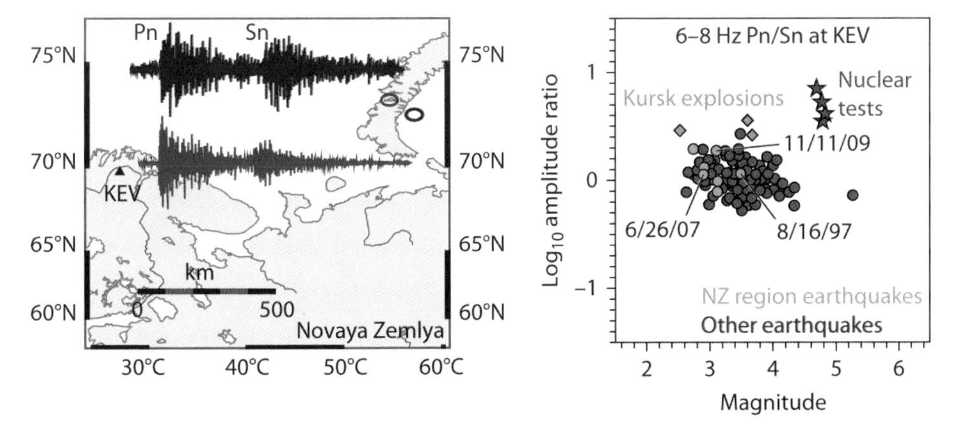

Ratios of high-frequency P to S waves discriminate (identify) Novaya Zemlya nuclear tests from earthquakes. Left-hand side compares 6–8 Hz (cycles per second) seismic waves at Kevo, Finland, for the 1997 Kara Sea earthquake (in blue) with a nuclear test in 1990 (in red). Right-hand side shows P/S values for five nuclear tests, nonnuclear explosions related to the sinking of the Kursk submarine, and earthquakes on and near Novaya Zemlya. The 1997 and more recent earthquakes in 2007 and 2009 are labeled. *Source*: National Academies Report, 2012.

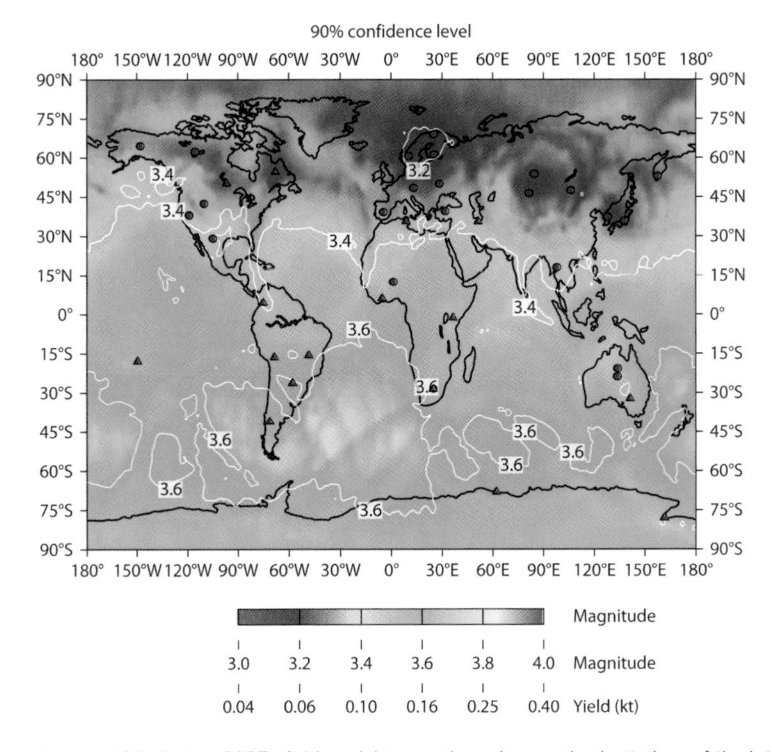

Detection capability in late 2007 of thirty-eight operating primary seismic stations of the International Monitoring System. Contours indicate the magnitude of the smallest event that would be detected with high likelihood (90 percent probability). Prepared by Kværna and Ringdal of NORSAR with yields added for National Academies Report of 2012.

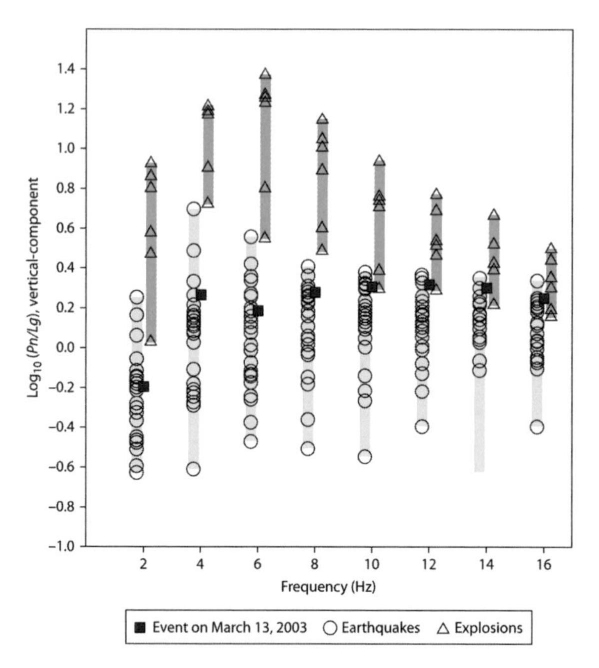

Measurements of high-frequency seismic waves for earthquakes and explosions near the Chinese test site at Lop Nor from 2000 to 2008. The log of the amplitude ratio of P to Lg seismic waves is shown on the vertical axis. Triangles denote earlier nuclear explosions at regional stations in blue. Yellow circles indicate earthquakes. Explosions have higher values on the vertical axis than earthquakes for frequencies of 4 and 8 Hz (cycles per second). *Source*: Kim, Richards, and Sykes, 2009.

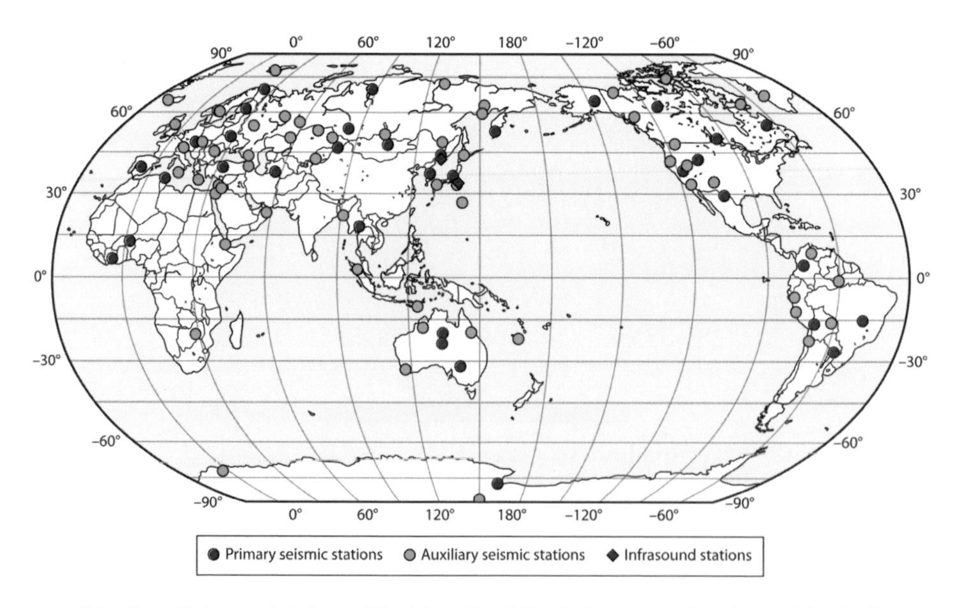

Seismic and infrasound stations of the International Monitoring System that detected the North Korean nuclear explosion of 2013. *Source*: Comprehensive Test Ban Treaty Organization, 2013.

11

RENEWED INTEREST IN A CTBT, THE OTA REPORT, AND THE GROUP OF SCIENTIFIC EXPERTS: 1979–1996

Much happened about nuclear testing during the fifteen years prior to the signing of the Comprehensive Nuclear Test Ban Treaty (CTBT) in September 1996. U.S. congressional committees expressed considerable interest in it. I wrote and testified extensively about the verification of a CTBT and the determination of yields of Soviet explosions. From 1986 to 1988 Congress's Office of Technology Assessment (OTA) conducted the first independent review of the determination of Soviet yields and how well a CTBT could be verified.

During my sabbatical leave from Columbia University from July 1981 to June 1982, I corresponded with Jack Evernden of the U.S. Geological Survey and Dennis Flanagan, the editor of *Scientific American*, about an article on the verification of a CTBT. Flanagan accepted a proposal by Evernden and me, and our article "The Verification of a Comprehensive Nuclear Test Ban," appeared as the lead article in their October 1982 issue. At that time *Scientific American* and its publisher, Gerard Piel, were known for including articles on various aspects of the nuclear arms race. Evernden and I exchanged drafts of materials for the article by airmail in early 1982 when I was at Clare Hall College of Cambridge University and he was working for the U.S. Geological Survey.

I find in rereading our 1982 article that it is still relevant today for educated, technical, and public policy audiences. It covers most of the issues related to the verification of a CTBT and came very close to forecasting the capabilities that now exist for using internal and external seismic networks to monitor the Soviet Union. Our subheading in 1982 stated, "Networks of seismic instruments could monitor a total test ban with high reliability. Even small clandestine explosions could be identified if extreme measures were taken to evade detection." We commented that the issues to be resolved were political, which I think still is the case.

Evernden, Cifuentes, and I followed up our 1982 paper the next year with a longer, more detailed analysis of monitoring the Soviet Union. We showed that its areas of thick salt deposits could be monitored with stations in the USSR down to yields smaller than one kiloton even if small evasive tests were contemplated using underground cavities.

Our 1982 article helped to stimulate renewed interest in a full test ban throughout the 1980s and 1990s. In 1986 Evernden, Archambeau, and I received the Public Service Award from the Federation of American Scientists for "Leadership in Applying Seismology to the Banning of Nuclear Tests, Creative in Utilizing Their Science, Effective in Educating Their Nation, Fearless and Tenacious in Struggles within the Bureaucracy."

NEGATIVE VIEWS ABOUT MONITORING A CTBT BY BACHE AND ALEWINE OF DARPA

Evernden and I invited Bache and Alewine of the Defense Advanced Research Projects Agency (DARPA) to give talks at a symposium titled "Verification of Nuclear Test Ban Treaties" which we organized for the American Geophysical Union in June 1983. This chapter discusses presentations at that symposium and conclusions in the OTA report of 1988 about verifying a full test ban. Determining yields of underground explosions and accusations of Soviet cheating on the Threshold Treaty were covered in the previous chapter.

In an unpublished 1983 manuscript "Monitoring a Comprehensive Test Ban Treaty," Bache and Alewine stated, "The conclusion is that serious technical problems remain to be solved to attain the capability to assert (with any confidence) that there is no clandestine program of a few events/year with yields up to a *few tens of kilotons* [my italics]." Bache criticized our *Scientific American* paper, stating, "In that article they state that the reliability of measures for the verification of a treaty banning explosions larger than one kiloton is no longer arguable. I assume that this must be an overstatement of their position, otherwise there wouldn't be any point to this symposium." In the 1980s Evernden and I claimed that two kilotons could be detected, much better than a few tens of kilotons.

To my knowledge, the U.S. government has never issued an unclassified statement about what yields constitute tests of military significance. Its absence has led opponents of a CTBT to change their positions over time about the importance of tests of various yields, no matter how small. What is too small, of course, involves policy as well as technical considerations.

Bache and Alewine are correct that a reliable regional method of identification is needed if the United States wants to identify events smaller than seismic magnitude m_b 4.0. That is, in fact, why Evernden, Cifuentes, and I made computations in 1983 for a network that included fifteen seismic stations at regional distances in the USSR—that is, between 0 and 1250 miles (0–2000 km)—as well as more distant external stations.

Bache and Alewine presented a table of magnitudes and yields for explosions at the Nevada Test Site in (1) competent water-saturated rocks and (2) dry porous materials like dry alluvium. In 1982 Evernden and I stated that the maximum thickness of dry alluvium in the USSR leads to an upper limit of two kilotons for that evasion scheme, as scientists in the military branch of the U.S. Geological Survey had stated years before. In addition, explosions in dry alluvium are likely to create craters at the earth's surface, which can be detected by satellites. Bache and Alewine made no mention of these difficulties in associating magnitudes of 4.0 and 4.5 with yields of 15 to 90 kilotons in alluvium.

Bache and Alewine also cited the difficulty of detecting and identifying seismic events in the few hours after a large earthquake. That scenario, which is called "hide-in-earthquake," has been resolved for a long time. The two subfigures of figure 11.1 show seismograms made in Norway for the same time. Both show clear signals from a large earthquake in the eastern Soviet Union. The lower subfigure, however, was filtered to better display high-frequency seismic waves and to deemphasize low frequencies. Unlike the standard record at the top, it clearly shows a large arrival from a very small explosion of about 0.5 kiloton in Eastern Kazakhstan following the signal from the large earthquake.

What about testing in a large nearby earthquake? Since earthquakes cannot be predicted except for time periods of a few decades, a potential

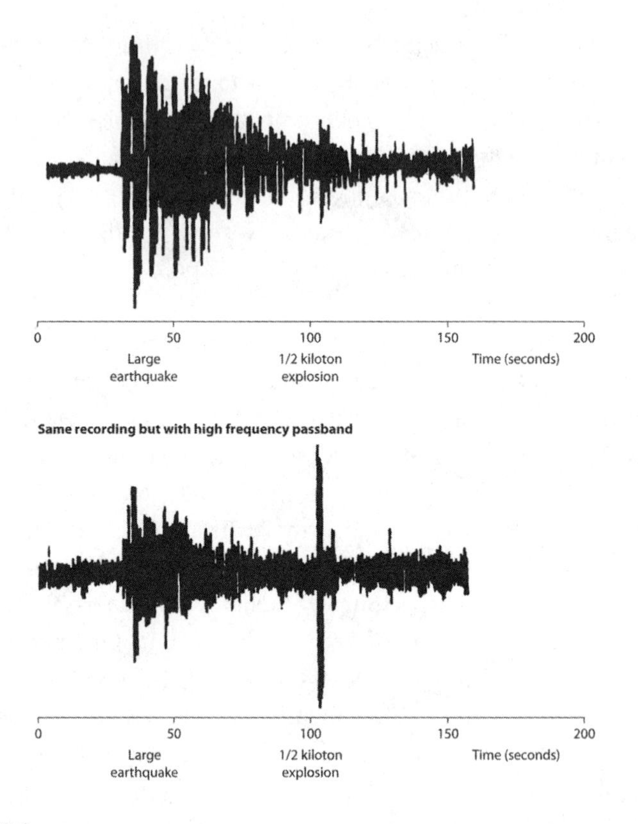

FIGURE 11.1

Seismograms from Norway for the same day and time in 1979. Time increases to the right from 0 to 200 seconds. Upper figure is a conventional recording of low-frequency seismic waves. Both figures show seismic waves starting at about 30 seconds from a large earthquake in the eastern Soviet Union. Seismic waves in the lower figure also include a large, impulsive P wave arriving just after 100 seconds from a very small explosion in Eastern Kazakhstan.

Source: Office of Technology Assessment, 1988.

violator would have to wait years to decades to set off an explosion in a large *nearby* earthquake. The violator would need to have a nuclear explosion ready to detonate at a moment's notice and to determine the location and size of the earthquake accurately within a few minutes. "Hide-in-earthquake" is not a realistic evasion scenario even for an explosion of a fraction of a kiloton.

The remarks by Bache and Alewine in 1983 and previously indicate that they and their office in DARPA did not take problem-solving approaches to test ban verification. Nevertheless, DARPA's mission going back to 1960 was to improve nuclear verification. DARPA does deserve credit for subsequently funding programs to analyze large volumes of seismic data and for promoting the installation of small seismic arrays.

SEISMIC MONITORING BY U.S. SCIENTISTS IN THE SOVIET UNION

In 2013 Frank von Hippel stated that he personally showed Figure 11.1, illustrating the detection of a very small Soviet explosion, to General Secretary Gorbachev in July 1986. Elected to that position in March 1985, Gorbachev and several of his more liberal scientific and technical advisers, such as Evgeny Velikhov, wanted to end the nuclear arms race. Gorbachev's first arms control initiative, on July 30, 1985, was to declare a unilateral moratorium on nuclear weapons tests for the remainder of that year—to be extended indefinitely if the United States reciprocated. Von Hippel states, "Reagan administration spokesmen argued that the Soviet Union had chosen to stop testing only after deploying a whole new generation of warheads. By contrast, they said, the US needed to test new warheads for the MX and Trident II missiles and for Edward Teller's antimissile x-ray lasers. . . . Moreover, they contended, a test moratorium could not be verified."

Von Hippel says that he met Velikhov at a conference in October 1985, and "Velikhov suggested that since the U.S. government was not interested in a mutually verified test moratorium, perhaps some non-governmental organization might be interested in verifying that the Soviets were not testing, even at low yields." The Natural Resources Defense Council (NRDC), a private environmental group in the United States, was interested.

In May 1986, NRDC signed an agreement with the Soviet Academy of Sciences to deploy three seismic stations near the nuclear test site in Eastern Kazakhstan. The stations, which were to be operated jointly and the data made available publicly, operated successfully for about a year. I was on NRDC's advisory committee from 1986 to 1988. Archambeau and several seismologists from UC San Diego set up and maintained those stations. NRDC also helped to set up three similar seismic stations around the Nevada Test Site, which involved U.S. and Soviet scientists. The stations gave important seismic information on tests set off by the other country. Most important, they helped to move the CTBT debate forward.

CONGRESS'S OFFICE OF TECHNOLOGY ASSESSMENT (OTA) ON VERIFICATION OF A FULL TEST BAN

The 1988 OTA report *Seismic Verification of Nuclear Testing Treaties* was the first serious analysis of test ban issues in the United States outside of the departments of Defense and Energy. It stated, "Verification—the process of confirming compliance and detecting violations if they occur—is therefore central to the value of any such treaty. Yet in the arena of arms control, the process of verification is political as well as technical. It is political because the degree of verification needed is based upon one's perception of the benefits of a treaty compared with one's perception of its disadvantages and the likelihood of violations." It involves an assessment of what is an acceptable level of risk and a decision as to what should constitute significant noncompliance. "Consequently, people with differing perspectives on the role of nuclear weapons in national security and on the motivations of Soviet leadership will differ on the level of verification required."

About half of the 1988 OTA report was devoted to monitoring the Soviet Union under either a full nuclear test ban or a low-yield threshold treaty. It stated, "A hypothetical seismic network with stations only *outside* the Soviet Union would be capable of detecting well-coupled explosions with yields below 1 kt anywhere within the Soviet Union and would be able to detect even smaller events in selected regions." It went on, "From a monitoring standpoint, stations within the Soviet Union are important [more] for improving identification capabilities than for further reduction of the already low detection threshold." The report cited the recording of high-frequency seismic waves by the three NRDC stations in Kazakhstan, but OTA had few other data on capabilities for monitoring within either Russia or China.

The OTA report was published just prior to the breakup of the Soviet Union. Many earthquakes once in the former Soviet Union, especially those in areas of thick salt deposits, are now located in independent countries. This considerably reduces the number of small earthquakes and potential nuclear explosions by the Russian Republic that must be monitored.

In a section titled "How Low Can We Go," the 1988 OTA report discussed various Soviet yield levels and how well they could be monitored.

Level 1—yields above 10 kt. Nuclear tests can be monitored with high confidence with external seismic networks and other national technical means [satellite images and other intelligence]. No method of evading a seismic network is credible.

Level 2—below 10 kt but above 1–2 kt. Demonstrating a capability to defeat credible evasion attempts would require seismic stations throughout the Soviet Union (especially in areas of thick salt deposits) and provisions in the treaty to handle chemical explosions. Expert opinion about the lowest yields that could reliably be monitored ranges from 1 kt to 10 kt.

Level 3—below 1–2 kt. The burden on the monitoring country would be much greater.

Level 4—comprehensive test ban. There will always be some threshold below which seismic monitoring cannot be accomplished with high certainty. The OTA report states that a comprehensive test ban treaty could still be considered adequately verifiable if it were determined that the advantages of such a treaty would outweigh the significance of any undetected testing below the monitoring threshold.

Experts from the weapons labs and the Department of Defense (DoD) were involved in the four panels convened by OTA from 1986 to 1988. It was not clear until the last minute if the DoD would permit Robert Zavadil, chief of the Evaluation Division, Directorate of Geophysics of AFTAC, to brief the OTA panels at the Secret level about classified network capabilities, identification, and yield determination. He briefed the OTA panels in an informative and straightforward way. Nevertheless, some officials of DoD (not Zavadil) criticized the OTA report soon after it was published in 1988.

OTA went on to conduct a separate study, *The Containment of Underground Nuclear Explosions*, which was published in 1989. Gregory van der Vink was the project director. I discuss containment, which involves assuring that radioactive products are not released into the atmosphere, in later chapters on international monitoring and evasion.

LETTER FROM BACHE TO ME ABOUT HIGH-FREQUENCY SEISMIC MONITORING

The American Geophysical Union asked me to write a six-page review on nuclear testing for the *U.S. National Report to the International Union of Geodesy and Geophysics*, which is published every four years. My paper "Underground Nuclear Explosions: Verifying Limits on Underground Testing, Yield Estimates, and Public Policy" was printed in *Reviews of Geophysics* in 1987.

Bache soon wrote to me (figure 11.2) concerning my statement that the paper "An Evaluation of Seismic Decoupling and Underground Nuclear Test Monitoring Using High-Frequency Seismic Data," by Evernden, Archambeau, and Edward Cranswick of the U.S. Geological Survey, probably was the outstanding article of the last four years published by U.S. scientists on test ban verification. I had made a general statement about the potential use of high-frequency seismic waves at regional distances in the Soviet Union. The paper by Evernden and colleagues represented an effort to move forward on the monitoring of explosions smaller than one to two kilotons, especially ones that might be detonated in an evasive manner. I think it was the start of an important process, not the end. The monitoring of high-frequency waves by seismic stations at distances up to 1250 miles (2000 km) within the Russian Republic, China, and surrounding countries became one of the outstanding contributions to nuclear verification in the following twenty-five years.

In his letter, Bache asked me if I had read their paper. Of course I had; undoubtedly he was being facetious. He went on to say that "this article goes beyond any reasonable bounds—and, of course, many of the seismological 'conclusions' are absurd." Then he stated, "Rather, I assume it reflects your opinion that the article is politically correct. That is a sort of

August 7, 1987

Professor Lynn R. Sykes
Lamont-Doherty Geological Observatory
Palisades, New York 10964

Dear Lynn,

I have just seen your article in the U.S. National Report to the IUGG. The length limitations make the review you undertook a very difficult task, and I compliment you on the excellent job you did. However, I must raise a flag in defense of some scientific standard in our politically battered field. The quote I find offensive is:

> The lengthy publication on high-frequency seismic propagation and its use in test ban monitoring by Evernden et al. (1986) is probably **the outstanding article of the last four years** published by U.S. scientists on test ban verification. [my emphasis added]

I suppose I must agree that the article is important in the sense that it has caused a lot of action (constructive and otherwise), but.... Have you read the article? Does it reach a scientific standard you consider appropriate? Would you pass a student who mistreats the scientific method as it is in this article? I recognize that seismology is an empirical science that tolerates a departure from rigorous standards of proof, but this article goes beyond any reasonable bounds -- and, of course, many of the seismological 'conclusions' are absurd.

Some of my younger staff who were upset by that quote (the young want to believe that scientific merit is the currency of greatest value) are urging me to send you technical material demonstrating the absurdity of material in this paper you set above all others. I would be happy to do so if I thought we were engaged in a technical debate, but that seems naive. Actually, you have studied the subject carefully, so I cannot believe your praise is based on technical grounds. Rather, I assume it reflects your opinion that the article is politically correct. That is sort of a 'red guard' approach to science, isn't it?

This is just one quote in an article that is otherwise a reasonable short review of a large technical subject with unavoidable political overtones. But you can see that it goes beyond bounds I am willing to tolerate quietly.

Sincerely,

Tom Bache

Thomas C. Bache
Manager
Geophysics Division

rec Aug 87

SAIC Science Applications International Corporation

Some things are beyond endurance . . .
In case you were hoping someone would say this -- here it is. I am contemplating ways to go public -- it is just a matter of time and priorities!

FIGURE 11.2

Letter from Thomas Bache, 1987.

a 'red guard' approach to science, isn't it? . . . But you can see that it goes beyond bounds I am willing to tolerate quietly." In a note attached to his letter, Bache wrote, "Some things are beyond endurance. . . ."

On July 8, 1987, Bache also wrote to van der Vink, the head of the OTA study on seismic verification, complaining about Evernden's

"Post-Meeting Calculations and Discussions." He stated, "From a technical perspective it seems that the Evernden & Archambeau work should collapse on its own weight, but I realize that it is difficult to discuss anything in this subject from a purely technical perspective." Bache took his opinions and political views as correct and those of others, like me, as tainted. I did not reply to him.

DARPA had previously set up a Center for Seismic Studies in Arlington, Virginia, which operated under contract to the consulting firm S-CUBED and employed several excellent seismologists. When Romney retired from DARPA, he moved to SAIC. The contract for operating the Center for Seismic Studies was soon up for renewal. Two consulting firms—S-CUBED and SAIC—bid to run it. Even though S-CUBED had much greater seismological expertise, the contract was awarded to SAIC. As far as I know, S-CUBED did not complain officially to DoD; complaining likely would have jeopardized their obtaining future DoD contracts. An irony is that S-CUBED was purchased by Maxwell Industries, which later did away with S-CUBED on the grounds that it was not making enough money. Scientists at S-CUBED then moved to SAIC.

CONGRESSIONAL ACTIONS TO HALT NUCLEAR TESTING

Congress moved on several fronts toward a Comprehensive Nuclear Test Ban Treaty (CTBT). In February 1986, the U.S. House of Representatives passed a joint resolution by a vote of 268 to 148 requesting that President Reagan resume negotiations with the USSR toward a Comprehensive Treaty and submit the Threshold and Peaceful Explosions treaties to the Senate for ratification. A similar proposal had passed the Senate by a vote of 77 to 22 in 1985. The House passed an amendment in August 1986 deleting funds for all U.S. tests in 1987 larger than one kiloton provided the Soviet Union would do likewise and would also accept a U.S. monitoring program. The House dropped its amendment prior to the Reykjavik summit when the Reagan administration agreed to submit the Threshold and Peaceful Nuclear Explosions treaties to the Senate for its advice and consent. The House voted again in May 1987 on a similar amendment. That November formal negotiations opened in Geneva on the limitation of nuclear tests.

In September 1992, the U.S. Senate, by an overwhelming vote, passed the Hatfield-Exon-Mitchell Nuclear Moratorium Amendment, which had three key elements. One set a deadline for the United States to stop testing. The second required a major scientific effort to ensure that we could maintain confidence in our nuclear weapons, absent actual nuclear tests. The third mandated that the next administration negotiate a CTBT no later than September 1996.

Just before the presidential election of 1992, President George H. W. Bush reluctantly decided not to veto a funding bill that included the moratorium amendment. It phased out U.S. nuclear testing except for fifteen tests, at most, if they were needed to deal with safety or reliability issues. Hazel O'Leary, President Clinton's energy secretary, stated that no tests of those types were required.

Nevertheless, some members of Congress and several of the directors of the nuclear weapons labs in the United States continued to argue that nuclear testing was needed to ensure that our existing stockpile of weapons would work if testing ceased, to develop new and safer weapons, to ensure that other countries would not cheat under a full test ban treaty, and to retain expertise in designing and maintaining nuclear weapons.

STRONG VIEWS BY KIDDER ABOUT QUESTIONABLE NEEDS FOR NUCLEAR TESTS

Ray Kidder, one of the most senior nuclear scientists at Livermore, argued that few U.S. nuclear explosions had been conducted to test the reliability of existing weapons a number of years after each weapon had been tested several times prior to its deployment. Later tests for reliability were so few that they were not a meaningful statistical measure of reliability. He said that most explosions, in fact, had been detonated to test new nuclear designs. A few were for so-called effects tests, which subjected electronic equipment, satellites, and delivery systems to the blast, heat, and radiation from a nearby nuclear explosion. Kidder clashed with several officials in the weapons labs over the need to continue testing existing weapons to confirm their reliability. Congress asked him to put his views in writing, which he did.

In 1985 Kidder published the percentage of U.S. nuclear weapons tests of various yields from 1980 through 1984. He argued that peaks in the number of tests near 5 to 20 and 150 kilotons could be taken as a measure of their perceived high military value. During that period, the military significance of tests below one kiloton was perceived to be low as judged from their small percentage (see Figure 12.6). The large numbers of tests near 150 kilotons resulted in part from testing strategic weapons at reduced yield that otherwise would have exceeded the limit of the Threshold Treaty. That limit of 150 kilotons was not important to the physics of weapons. Yields between 5 and 20 kilotons, however, were very important because they involved testing the primary (fission) triggers for thermonuclear weapons and testing at partial yield the ignition of the fusion (secondary) stage of weapons.

MY PUBLIC INVOLVEMENT WITH TEST BAN ISSUES

I was busy in the 1980s and 1990s on test ban issues and other aspects of the control of nuclear weapons. I co-taught a course on the nuclear arms race for undergraduates in 1984 and 1985. While it drew only twenty-five students per year, it attracted very bright undergrads from Columbia College and graduate students from the School of International and Public Affairs. The latter wanted more technical background on the arms race. I was a member of the Columbia University Seminar on Arms Control (for faculty and invited guests) from 1984 to 1996. In 1987 Paul Richards and I co-taught a Seismology Seminar course at Lamont on the "Verification of Nuclear Test Ban Issues."

From 1988 through 2000, I spoke in public on several occasions about the verification of nuclear testing. I participated in the Belmont Conference on Nuclear Test Ban Policy in 1988 and was an invited speaker at the Princeton Symposium on Non-Proliferation and Nuclear Testing in 1992. On May 31, 2000, I co-organized a second symposium for the American Geophysical Union on the Verification of the Comprehensive Nuclear Test Ban Treaty. I was a member of the board of directors of the Federation of American Scientists from 2000 to 2003 and was on their test ban panel.

INTERNATIONAL SEISMIC MONITORING EFFORTS

Starting in 1976, much happened at the international level as well as in the U.S. government on developing improved and more sophisticated systems of seismic monitoring of nuclear tests, including a rapid international exchange of data. In their 2009 book *Nuclear Test Ban: Converting Political Visions to Reality*, Ola Dahlman of the National Defense Research Institute of Sweden and his colleagues describe more about those international developments from 1976 until the signing of the Comprehensive Test Ban Treaty in 1996.

In 1976 the UN's Conference of the Committee on Disarmament (CCD) established "an Ad-Hoc Group of Government-appointed experts to consider and report on international co-operative measures to detect and identify seismic events, so as to facilitate the monitoring of a comprehensive test ban." It was referred to as the Group of Scientific Experts (GSE) and reported to the CCD and its successor, the Conference on Disarmament (CD).

Experts from many countries spent huge amounts of time each year on these endeavors, indicative of the importance most countries attached to halting testing and the development of more sophisticated nuclear weapons. China, France, Norway, Russia, Sweden, the UK, and the United States had long-standing programs in nuclear verification. Those governments picked some of their own scientists to represent them on the GSE. The U.S. representatives, who were appointed by the government, tended to be more conservative than many of us in the U.S. arms control community. Seismologists from small countries became involved in the technical details of monitoring through their participation in the GSE.

The GSE worked to design a global seismic verification system with rapid exchange of data, initially using the telecommunications system of the World Meteorological Organization. Fortunately, seismology and meteorology had long traditions of exchanging data globally on earthquakes and weather. The GSE proposed to collect information at special international data centers.

The first large-scale test, called GSETT-1 (Group of Scientific Experts Technical Test-1), took place for two months in late 1984. It involved the daily exchange from seventy-five stations in thirty-seven countries of very basic seismic parameters, such as the arrival times and amplitudes of P waves from earthquakes and explosions. Prototype International Data Centers, which prepared preliminary lists and final bulletins of seismic events, operated in Moscow, Stockholm, and Arlington, Virginia (near Washington, DC). The DARPA Center for Seismic Studies, which was one of these, was also the U.S. prototype National Data Center. Ann Kerr of DARPA was the U.S. coordinator for GSETT-1.

The GSE held a technical workshop on the design and function of an international data center in October 1987. The international group concluded after GSETT-1 that a future global system should involve seismic data that were recorded digitally. It conducted the second large test called GSETT-2 in 1991. By then many of the stations recorded data digitally, and more seismic array stations were operating.

The group conducted its third and last large-scale test, GSETT-3, starting full-scale operation on January 1, 1995, with data from sixty countries. It served as a prototype for the verification system adopted during the negotiations for the CTBT from 1995 to 1996. A single international data center located in Arlington, Virginia, received seismic waveform data in digital form. Waveforms contain much greater information than the exchange of simple parameters like the arrival times of P waves during GSETT-1. The center employed a staff of about fifty people from various countries.

The GSETT tests continued through the buildup of facilities for the Comprehensive Nuclear Test Ban Treaty Organization (CTBTO) in Vienna, Austria. The center in Arlington closed its prototype International Data Center (IDC) function in March 2000. Data from stations making up what was called the International Monitoring System (IMS), a part of the CTBTO, shifted from Arlington to Vienna. The computers and data handling in Vienna were similar to those in Arlington.

The GSE and its tests fostered international cooperation on tangible tasks. It also provided aid to developing countries in establishing modern seismic stations and arrays, training personnel and familiarizing diplomats and scientists with knowledge about nuclear verification. It showed that an international monitoring system and data transmission were realities and not abstract concepts. DARPA provided substantial funds for the operation of the data center in Arlington and for much more powerful computers and programs to analyze seismic waveforms.

INITIATIVES AND FUNDING BY DARPA

DARPA formed a Research and Systems Development Initiative, which they announced to industry in November 1987. In addition to the operation of the prototype national and international data centers, the initiative proposed a next generation capability for integrated data processing, high-speed global communication (mainly by satellite), authentication of data, long-term storage of what were then vast amounts of data, and a research test bed for examining new concepts and for signal analyses. The operation of a state-of-the-art center in Vienna, rapid communication of data, processing of full waveform data, and rapid distribution of data and results to national data centers would not have occurred so quickly without DARPA's initiatives and funding.

In contrast, the U.S. national center was a stand-alone facility. Its database was available only to designated users in the United States. The national center included work on yield estimation, which the international center did not, and analysis of classified U.S. data.

DARPA's method of operation was to contract the development of seismic instruments, arrays, satellites, communications, data analysis, and research to private industry and occasionally to universities. It employed relatively few people itself in nuclear monitoring. In contrast, the weapons labs did many similar tasks in-house with their own personnel. Alewine was the director of DARPA's Nuclear Monitoring Office from 1980 to 1996, when he became deputy assistant secretary of defense for nuclear treaties and his office was transferred to the Office of the Secretary of Defense. He is now retired.

DARPA officials stymied consideration of a full test ban by the U.S. government for decades on the grounds that it could not be verified and argued for twenty years that the Soviet Union was testing weapons with yields larger than those permitted under the TTBT. By 1987, however, officials in DARPA became very involved in helping to create modern instrumentation and data transmission facilities for monitoring a full test ban. That decision was likely driven by four factors: (1) increased political pressures in the United States and internationally for a CTBT, (2) recognition that DARPA could no longer continue arguing about the yields of Russian explosions, (3) acknowledgment that Russia was testing within the 150-kiloton limit of the Threshold Treaty, and (4) wanting to show that DARPA could draw upon and use its long technical and scientific expertise in defense issues, as it had done it developing the Internet.

DARPA's influence on U.S. policies about nuclear monitoring and a CTBT was profound. In my estimation, much of their long influence was negative and based on poor science. Officials well below the level of the secretary of defense generated most of DARPA's policies about testing, which were then passed up to higher authorities. The U.S. Defense Department had employed "red teams" to criticize some proposed policies, but to my knowledge it did not do so for test ban issues. More independent analysis within the executive branch of the U.S. government and stronger congressional oversight might have helped to resolve test ban controversies much sooner. The 1988 report *Seismic Verification of Nuclear Testing Treaties* by Congress's Office of Technology Assessment helped to constrain test ban issues technically, but it occurred more than thirty-five years after the first hydrogen bomb explosion.

12

DEALING WITH "PROBLEM" OR "ANOMALOUS" EVENTS IN THE USSR AND RUSSIAN REPUBLIC: 1972–2009

A debate started in 1972 and lasted until 2009 about the identification of so-called problem or anomalous seismic events that some people believed to be difficult to identify. Occasional problem events such as these, which are likely to occur about once a year, require more detailed examination than most earthquakes do. The following examples illustrate that with special studies, all of them through at least 2009 can be identified as earthquakes. I pay particular attention in this chapter to very small seismic events near the Russian test site on Novaya Zemlya, several of which the U.S. government claimed were either nuclear explosions or possible explosions.

U.S. PAPER SUBMITTED TO THE UN IN 1972 ON ANOMALOUS SEISMIC EVENTS

In August 1972 the U.S. government presented a working paper, *A Review of Current Progress and Problems in Seismic Verification*, at the Conference of the Committee on Disarmament (CCD) of the United Nations. It listed twenty-five "anomalous" or "problem" seismic events that presented difficulties in identifying underground explosions as opposed to earthquakes using the difference between long- and short-period seismic waves by the Ms-m_b technique, discussed in chapter 6. The U.S. working paper described them as "false alarms," identifying them as likely earthquakes that lay either in or close to the nuclear explosion portion on an Ms-m_b diagram.

Although no author is listed on the 1972 working paper, officials in the Department of Defense likely wrote the document with data processing being performed by a consulting firm. Like many other official U.S. reports, it emphasized many problems in seismic verification and indicated that the Ms-m_b technique might well fail to identify future seismic events. Other than stating the need for more research, it did not take a problem-solving approach. These reports typically convey the sense that the "glass is half empty." The report's implication clearly was that seismic identification was not good enough to justify the United States' entering into a full test ban treaty.

Although considerable information from other seismic stations was readily available, most of the original magnitudes in the 1972 report were determined using data from only one or two stations. Four years later, in 1976, Lamont graduate student Robert Tathum, research scientist Donald Forsyth (now at Brown University), and I made a thorough study of each of the twenty-five events included in the 1972 report, using available unclassified seismic data. We increased the number of good m_b determinations from the original 83 in the 1972 report to 242. We published our results in a peer-reviewed scientific journal.

We showed that with additional data about half of those twenty-five events had Ms-m_b values that put them clearly in the earthquake population—that is, those events ceased to be "anomalous." Several of the remaining problem events occurred at depths of 15 to 30 miles (25 to 50 km). They could be identified easily because nuclear explosions have not been set off at depths greater than about 1.2 miles (2 km). The greatest depths of petroleum and other wells do not exceed 7 miles (11 km). The sole exception was a single super-deep hole drilled to a depth of about 10 miles (15 km) for scientific purposes by the Soviet Union in hard, ancient rock at a huge cost. It was unstable and shrank in size at those depths and would therefore have been unsuitable for a nuclear test.

This left one unresolved sequence of problem events in a small, very mountainous area of Tibet that occurred during a three-month period in 1968 and 1969, and some events in this swarm were larger than magnitude 5. Nevertheless, at several stations the first motions of their P waves clearly were downward, indicating they were earthquakes and not explosions. Other events in the swarm had waveforms almost identical to those of the larger events in the series, which led us to conclude that all of them were earthquakes. Even in 1974, we were able to identify all of the claimed twenty-five "anomalous" events as earthquakes. Today much better seismic data exist from stations within Asia, so that events like these can be readily identified.

Understandably, the United States and other countries do not want to miss even a single nuclear explosion, nor do they want to incorrectly identify an earthquake as an explosion. This is a big job, especially with small explosions, because the number of earthquakes increases by a factor of about eight to ten when seismic magnitude decreases by one unit. Additional methods besides the standard Ms-m_b technique should be applied to distinguish earthquakes from explosions. For the few "problem" events, I recommend a thorough, unbiased, and problem-solving approach using a variety of seismic methods developed by experts with a long tradition of work in seismic verification. U.S. officials who reviewed the 1972 report prior to its submission to the UN should have caught the low number and incomplete nature of the magnitudes used.

SMALL SEISMIC EVENTS NEAR THE RUSSIAN ARCTIC TEST SITE

During the 1980s and 1990s and into the following decade, some analysts in the United States claimed that several seismic events were either small nuclear explosions or possible explosions. A number of these accusations were released to the press, particularly to the *Washington Times*. Some came at critical times, either during or soon after the negotiations for the Comprehensive Test Ban Treaty (CTBT) in 1995 and 1996 or the U.S. Senate's debate on the treaty in 1999. It is important to note that all of these "problem" events were identified as earthquakes by either British, Norwegian, Livermore, or Lamont seismologists.

In 1987 Ryall and others at DARPA's Center for Seismic Studies were not able to identify a small seismic event on or near Novaya Zemlya on August 1, 1986 (Figure 12.1), as being either an explosion or an earthquake. The large uncertainty in the computed initial location included a small area of land to the southeast of the Soviet arctic test site, but not the test

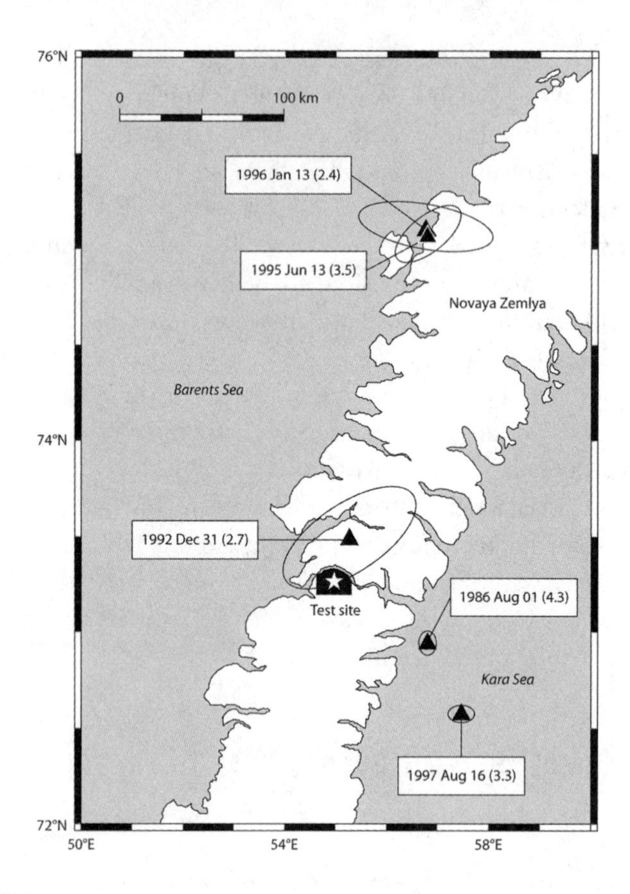

FIGURE 12.1

Locations (triangles) and their uncertainties (open circles) of five small earthquakes on and near Novaya Zemlya from 1984 to 1997. The seismic magnitude m_b is shown in parenthesis following the date of each event.

Source: Sykes, 1997.

site itself. Their results were widely distributed, and several people indicated that similar problem events were an obstacle to achieving a full test ban treaty.

Novaya Zemlya and its surrounding Kara and eastern Barents seas are not very active for moderate to small earthquakes (figures 12.1 and 12.2). Novaya Zemlya is about 625 miles (1000 km) long, similar in size to California. A 200 by 200 mile (300 by 300 km) region centered on the

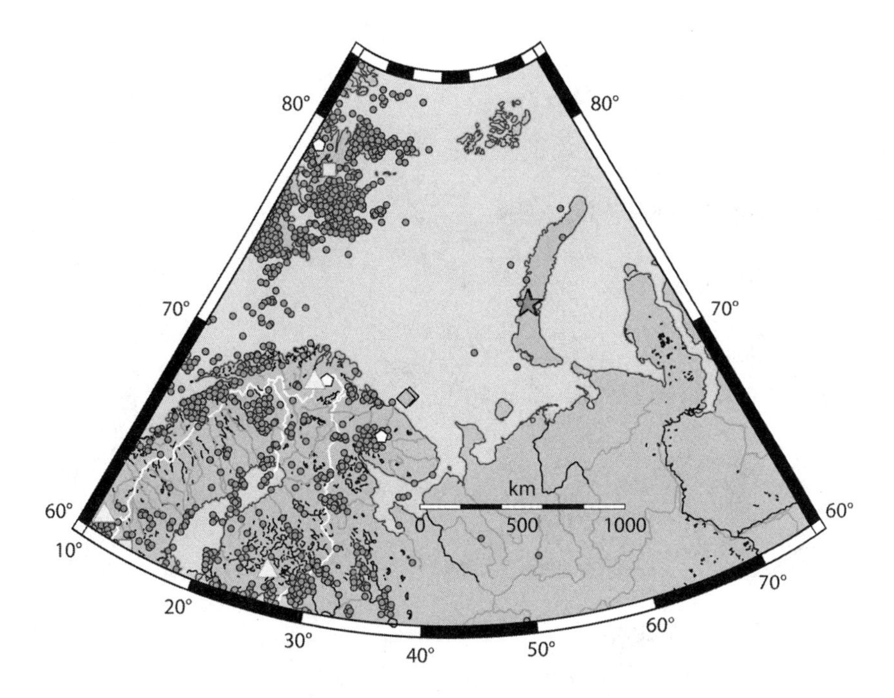

FIGURE 12.2

Seismic events and stations in the vicinity of the main Russian arctic test site at Novaya Zemlya. Stars show the locations of nuclear tests since 1977. Small circles indicate seismic events from 1999 to 2009 of magnitude greater than 2.0. The primary stations of the International Monitoring Service (triangles), the auxiliary station in Sweden (square), and three other publicly available stations (pentagons) are shown in white. The diamond locates the Kursk submarine disaster of 2000. Many of the events in Scandinavia, Finland, and mainland Russia were small mine blasts.

Source: National Academies Report, 2012.

main test site is less active than a similar area centered on New York City. Both are *intraplate* regions; that is, they are located inside a tectonic plate rather than along an active plate boundary.

The main Novaya Zemlya test site is located well north of the Arctic Circle, at 73 degrees north latitude, farther north than the northernmost point in Alaska. It is a frigid, largely mountainous region where polar bears are much more numerous than humans (figure 12.3). A glacier covers much of the northern island of the two parts of Novaya Zemlya.

FIGURE 12.3

Polar bear in front of the officers' club at Novaya Zemlya.

Photo courtesy of Paul Richards, 2004.

In 1989 Marshall and two colleagues from the British verification group made an extensive study of the 1986 seismic event and published it in a peer-reviewed journal. They examined more data and performed more identification tests than Ryall and others had done. They concluded the event was an earthquake based on the ratio Ms/m_b and a focal mechanism solution. Their uncertainty in location, which is shown in figure 12.1, placed it entirely beneath the Kara Sea. Their identification of the seismic waves P, pP, and sP on records from several stations at large distances allowed them to determine its depth as about 15 miles (24 km), another indication it was an earthquake. The paths those waves travel through the Earth are illustrated in figure 3.1. The thorough work of Marshall and colleagues convinced the nuclear verification community that the event clearly was an earthquake even though its magnitude m_b was small, only 4.3. If it had been a standard, well-coupled nuclear explosion, its yield would have been about one kiloton.

Detection, location, and identification of seismic events in and near the Soviet arctic test site improved after the early 1990s with the installation and operation of powerful seismic arrays (a system of linked seismometers arranged in a regular geometric pattern) in northern and southern Norway, southern Finland, and to the north of Norway in Spitzbergen (figure 12.4). Each array provides an estimate not only of the distance to a seismic event but also of its azimuth (or direction). An array is like either a spotlight that shines brightly only in a narrow beam or a directional microphone that is sensitive to sound from one azimuth. A single station, on the other hand, is akin to a light bulb that shines equally in all directions.

The four seismic arrays continuously monitor the Novaya Zemlya test site with an excellent detection capability down to magnitude 2.2 to 2.5

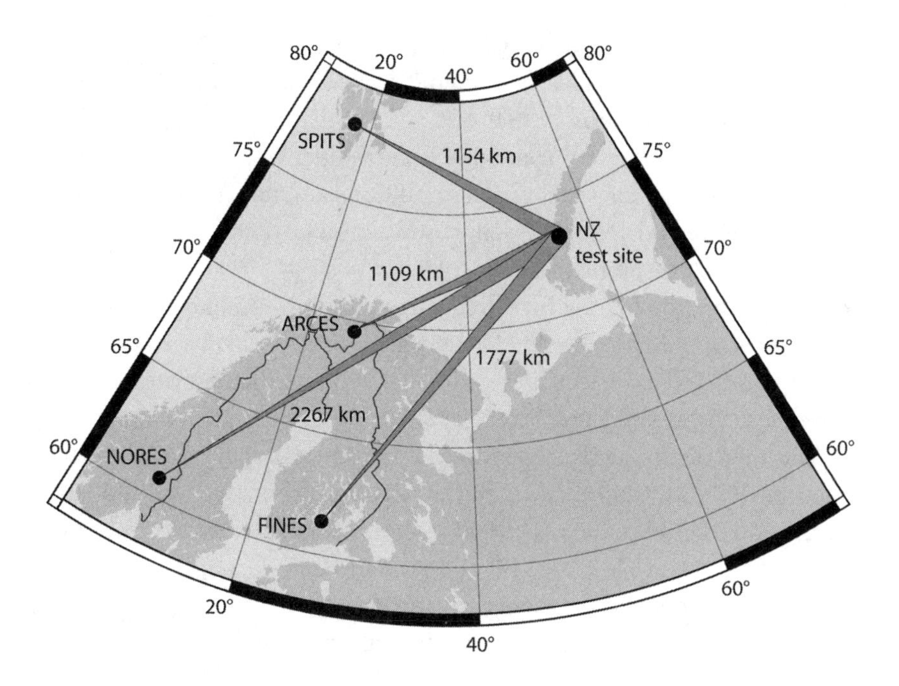

FIGURE 12.4

Locations of four seismic arrays that continuously monitor the Russian test site on Novaya Zemlya (NZ).

Source: Kværna, unpublished figure 2016.

(5 to 15 tons = 0.005 to 0.015 kilotons). These capabilities today are made possible by the successful operation of those arrays and decades of work by Norwegian seismologists. The Russian test site at Novaya Zemlya is one of the world's best-monitored places of high concern to the United States.

As the negotiations for the Nonproliferation and Comprehensive Test Ban treaties commenced in the 1990s, any small seismic events that were located on or near the two islands of Novaya Zemlya were leaked to the media as possible or likely Russian nuclear explosions. Four small seismic events from 1986 to 1997 were cited by the U.S. Defense Department, including one by the secretary of defense, as being of either suspicious or uncertain origin.

The event shown in figure 12.1 of magnitude 2.7 on December 31, 1992, took place near the main Russian test site. Since it occurred on New Year's Eve, some Americans and Russians have said most people at the test site would have been too drunk to conduct a nuclear explosion and Soviet authorities would not have permitted it to be conducted on New Year's Eve. It is important to note that of the forty-two underground nuclear explosions on Novaya Zemlya, none occurred between December 4 and May 7, times of harsh arctic winter conditions and no daylight. Hence, a test on New Year's Eve seems highly unlikely. Nevertheless, skeptics drew attention to it as a problem event.

Fortunately, the ratio of high-frequency P to S waves, like those shown in figure 12.5, also indicates the 1992 event was a very small earthquake. If it had been a well-coupled explosion in hard rock at that test site, which it was not, its yield would have been very small, about 0.025 kilotons (25 tons).

The yield of another small seismic event, on January 13, 1996, near the coast of the northern island of Novaya Zemlya (figure 12.1), would have been about 0.013 kilotons (13 tons) if it were an explosion. Even if it or the earthquake of New Year's Eve 1992 had been decoupled (muffled) explosions detonated in a cavity in hard rock, their yields likely would have been smaller than one kiloton. Leaking stories to the press about the smallest detected event, such as that of 1996, clearly is making a mountain out of a molehill—especially one such as this located 125 miles (200 km) from the Russian test site. Salt, which is the easiest common rock in which to form large cavities, is not present on Novaya Zemlya in any appreciable thickness.

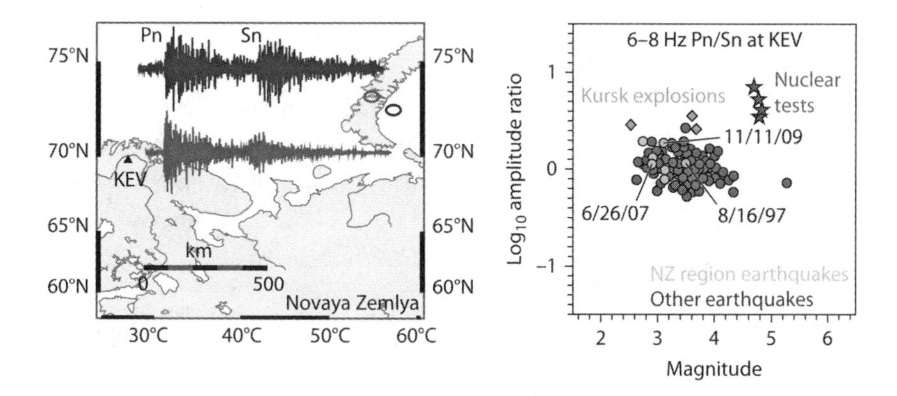

FIGURE 12.5

Ratios of high-frequency P to S waves discriminate (identify) Novaya Zemlya nuclear tests from earthquakes. Left-hand side compares 6–8 Hz (cycles per second) seismic waves at Kevo, Finland, for the 1997 Kara Sea earthquake (above) with a nuclear test in 1990 (below). Right-hand side shows P/S values for five nuclear tests, nonnuclear explosions related to the sinking of the Kursk submarine, and earthquakes on and near Novaya Zemlya. The 1997 and more recent earthquakes in 2007 and 2009 are labeled.

Source: National Academies Report, 2012.

U.S. CLAIMS THAT THE 1997 EARTHQUAKE WAS A RUSSIAN NUCLEAR EXPLOSION

A great governmental furor and media attention in the United States focused on a small seismic event in the Kara Sea on August 16, 1997. Officials in Moscow maintained that the event was a small earthquake in the Kara Sea to the east of Novaya Zemlya. Nevertheless, on August 28, 1997, the *Washington Times* carried a lead story by B. Gertz titled "Russia Suspected of Nuclear Testing." The *Washington Post* and the *New York Times* published similar stories the next day, the latter with the headline "U.S. Suspects Russia Set Off Nuclear Test." Remarks quoted in the press, such as "This one certainly had characteristics that at least would lead some to believe that there had been an explosion that caused the event," emphasized the likelihood of a clandestine nuclear explosion. Buried in the body of the text, however, was a statement that whether it was a nuclear explosion or a small earthquake was still in doubt.

This was the first of several allegations by U.S. officials of cheating by Russia after the UN General Assembly overwhelmingly adopted the Comprehensive Nuclear Test Ban Treaty (CTBT) in September 1996. The widely publicized allegation about the August 16 event and a formal protest by the United States to Russia came at a very sensitive moment, just as President Clinton, who appeared at the UN on September 22, 1997, reported that his administration would submit the treaty, which he had signed, to the U.S. Senate for its advice and consent.

Russia reported that a chemical explosion had been conducted two days earlier, on August 14, 1997, to test the compression of plutonium. They stated it was a so-called hydrodynamic test that released no nuclear energy. The United States had conducted and announced hydrodynamic tests in previous months at the Nevada Test Site. Hydrodynamic tests are permitted under the CTBT. Apparently, an earthquake two days after August 14 led officials in the United States to jump the gun, believing that the seismic event of August 16 was a small nuclear test.

By the time the story appeared in the *Washington Times*, twelve days after August 16 and at the start of the Labor Day holidays, a strong consensus had developed among seismological experts in the United States, the UK, Norway, and Canada that the event was, in fact, a small earthquake with a magnitude of 3.3 (figure 12.1). If it had been a well-coupled nuclear explosion, its yield would have been about 0.1 kiloton (100 tons). In fact, it was well located beneath the waters of the Kara Sea about 60 miles (100 km) southeast of the Russian test site and farther away from it than the 1986 earthquake.

Nonetheless, nearly two months after the 1997 event, several top U.S. officials were either unaware of this scientific consensus or chose to ignore it, maintaining that the nature of the seismic event was ambiguous. A quick initial location, incorrectly made by U.S. agencies, apparently was passed up chains of command in the government even though it became obvious just one day later that the event was located well at sea. Clearly, revised estimates need to be made and communicated to high officials in a timely manner as more critical data become available, particularly when an event is or may be of potential concern to national security.

High officials and some government agencies in the United States had difficulty admitting they were mistaken—that the seismic event was merely a small earthquake and initial leaks to the press were incorrect. When the correct information was finally released officially months later, the event, once a front-page claim, became a small back-page story.

Both the preliminary International Data Center (IDC) of the International Monitoring System (IMS), then situated in Arlington, Virginia, and the U.S. National Data Center played key roles in recording and analyzing seismic data on August 16, 1997. Under the terms of the Comprehensive Test Ban Treaty, the IDC and its successor in Vienna are not allowed to identify "problem" events as earthquakes, nuclear explosions, or chemical explosions. Those tasks as well as yield estimations are reserved to national CTBT authorities. The Air Force Technical Applications Center (AFTAC) operates the data center for the United States, but its results are classified.

Seismic stations of the International Monitoring Service and other stations in northern Europe (figures 12.2 and 12.4), except for the sensitive seismic array ARCESS in northern Norway, recorded the 1997 event. Although ARCESS was being repaired following a power surge, data from other seismic stations in the region were sufficient to locate and identify the event. The Finnish standard station KEVO near ARCESS, while not part of the IMS, recorded the earthquake as it had many previous nuclear explosions at Novaya Zemlya. KEVO's data were readily and quickly available over the Internet from the U.S. consortium for seismology called IRIS.

This illustrates the need to be ready to analyze available data rapidly from many stations for a "problem" event. When data from KEVO and several other stations in Scandinavia became available the next day, it was clear that the event was in fact in the Kara Sea, not on land (figure 12.1).

Many U.S. seismologists like me learned about the 1997 seismic event from press reports twelve days after it occurred. Several of us at Lamont, including Paul Richards, Won-Young Kim, and I, immediately examined seismic data from the event. We found that the use of high-frequency P and S waves provided clear additional evidence that the event of August 16, 1997, was an earthquake. It falls in the earthquake population of Figure 12.5 and not in that of nuclear explosions at Novaya Zemlya. In addition, Norwegian seismologists identified a very small aftershock of that earthquake by cross-correlating the signals from the two events.

The 1986 and 1997 earthquakes occurred near sites of Soviet nuclear waste disposal that would not be auspicious sites for conducting offshore nuclear explosions. (The occurrence of the small earthquakes there is not related to the disposal of wastes.) In addition, rocks in the area are very old, and it is not a site of present or recent volcanic activity.

On September 13, 1997, John Diamond of the Associated Press, under a headline "Study Supports Russia Test Denial," wrote, "New findings in a secret Air Force study indicate that a tremor thought to have been a Russian nuclear test occurred underwater, pointing to the likelihood it was an earthquake."

Jeffrey Smith, a staff reporter on national security for the *Washington Post*, wrote an extensive front-page article on October 20 titled "U.S. Officials Acted Hastily in Nuclear Test Accusation: CIA Hesitates to Call Russian 'Event' a Quake." Smith wrote, "A high-priority classified alert issued by the CIA on Aug. 18 quickly caught the eye of senior policy makers. The bulletin came from the government's Nuclear Test Intelligence Committee, an interagency scientific group, and said that Russia probably had conducted a nuclear test two days earlier on an island near the Arctic Circle. Officials at the National Security Council swung into action, convening an interagency meeting two days later and ordering a full-court press to collect an explanation from Moscow."

Jeffrey Smith reported that Harold Smith, assistant to the secretary of defense, said that other scientists at the Pentagon shared his belief that the initial CIA report was wrong. He quoted Harold Smith: "I personally think it was an earthquake. We now know that they would have been well advised to wait until they had more data and could reach an accurate conclusion." The article quoted my colleague Richards: "Not only was there a mistake made, but there was no effort to retract it." Smith also quoted Eugene Herrin, who for the previous fifteen years had chaired the military's principal seismological advisory panel: "somebody jumped the gun. Based on what I know, it was not an ambiguous event. . . . It's an earthquake."

I wrote a twelve-page article on the earthquake for the November/December 1997 issue of the Federation of American Scientists' *F.A.S. Public Interest Report*. I spoke to Jeffrey Smith about my forthcoming article, and he was able to obtain information from sources in the U.S. government to which I was not privy. Smith's article appeared on the same day I participated in a news conference held in Washington, DC, related to the August 16 event that was sponsored by the Coalition to Reduce Nuclear Danger.

One of the important documents I furnished to Jeffrey Smith was a copy of a fax sent by Frode Ringdal of Norway to Alewine of the U.S. Defense Department. In it Ringdal clearly put the event in the Kara Sea more than 60 miles (100 km) from the Russian test site. I realized that the date of the receipt stamp and fax numbers were still on the top of my copy of the fax. Alewine's office received it a week before Gertz quoted him in the *Washington Times* indicating that the event was a possible explosion. This vital information was a "smoking gun" that Smith used to get more material for his story. He did an excellent job obtaining information from many sources for his front-page story.

Jeffrey Smith went on to state on October 20, "But the nuclear intelligence committee, which the CIA chairs, did not formally begin backpedaling until two weeks after the event, causing one official to describe it as "the last to join the crowd." Articles of October 21, 1997, in the *Washington Post* and the *New York Times* indicated that the CIA was involved; no names were mentioned, but several people have told me independently that Larry Turnbull, a known hawk, was a chief player. He had stated at a previous international meeting at Princeton that he worked for the CIA. Hence, he was not working undercover.

Robert Bell, director of defense programs at the U.S. National Security Council, also participated in the news conference on October 20. He stated that considerable activity was observed at the Russian test site on August 14, two days before the seismic event. He also said that no seismic waves were detected on August 14 from Novaya Zemlya and that a plane sent out found no indications of radioactive release. Interestingly, on October 20, two whole months later, Bell still seemed uncertain about the nature of the seismic event of August 16.

A CIA press release on November 4, 1997, stated that "a seismic event occurred on August 16, 1997, in the Kara Sea. That seismic event was almost certainly not associated with the activities at Novaya Zemlya and was not nuclear. However, from the seismic data, experts cannot say with certainty whether the Kara Sea event was an explosion or an earthquake." Nonetheless, many experts in the United States and abroad had concluded weeks before that it was an earthquake. While the CIA press release goes some distance in saying the event was not nuclear, no mention was made publicly that U.S. officials had earlier made a rush to judgment. In my opinion, the experts cited in the 1997 CIA press release should have been fired.

The early press releases about the seismic event of August 16, 1997, led some conservative commentators and organizations to continue to dig in their heels about its being a nuclear explosion long after a scientific consensus emerged that it was an earthquake. For example, Jeffrey Smith's article stated, "Frank Gafney, director of the Center for Security Policy, said evidence his group is gathering bolsters the case that the tremor resulted from a nuclear test." Gafney, much like Richard Perle, frequently opposed various arms control treaties and proposals. Neither he nor Perle is a scientist, but each often has claimed to have inside information.

The Report on CTBT Technical Issues by the U.S. National Academies in 2012 "drew three lessons from the handling of the 1997 event: (1) use all available data for accurate estimates of location and event characterization, (2) avoid the use of only a narrow range of azimuths such as those to southern Norway and Finland, and (3) provide a mechanism for new information to be updated to policy makers as it becomes available for occasional 'problem' events of this type." Hopefully, these lessons have been learned, as small seismic events on Novaya Zemlya in 2007 and 2009 did not lead to leaks of false assessments to the media. Their high-frequency P to S wave ratios identified them as earthquakes (figure 12.5). Norwegian seismologists located the magnitude 2.8 earthquake of June 26, 2007, within 30 miles (50 km) of the main site where Russia had previously tested.

A visiting high-level Russian scientist told me later that Boris Yeltsin, president of the Russian Federation, was on vacation when the U.S. press reported the event of 1997 as a likely nuclear test. Yeltsin thought the head of the Russian Ministry of Atomic Energy had deceived him and that the agency had, in fact, conducted a nuclear test. He bawled out the minister over the telephone. Knowledgeable Russian geophysicists quickly conveyed to Yeltsin that the event was a small earthquake in the Kara Sea. My source told me this would have been a much more serious incident between the United States and the USSR if it had happened during the height of the Cold War.

While press reports about small seismic events on and near Novaya Zemlya largely ceased after 1997, some stories continued to be published in the *Washington Times* about work at the Russian test site and about Chinese earthquakes being nuclear explosions at or near their Lop Nor test site. Bill Gertz, who had previously leaked incorrect information about Russian seismic events in the 1990s, wrote a story on September 24, 1998, headlined "Blast in Arctic, Satellite Shows." He wrote, "Vehicle activity photographed recently by a U.S. spy satellite indicates Russia is preparing to set off an underground blast at a remote Arctic nuclear-testing site, *The Washington Times* has learned."

Gertz coauthored a similar story on June 18, 1999, under the headline "Inside the Ring: Small Nuke Test?" which said, "China set off a small nuclear-related blast over the weekend—days before U.S. Undersecretary of State Thomas Pickering arrived in Beijing to deliver the latest American apology, according to Pentagon intelligence sources." Daryl Kimball of the Arms Control Association emailed me, "My sources tell me that: the report is based on satellite imagery only; that there is no seismic signature for this event from the official U.S. seismic stations (which are capable of detection to very low levels around Lop Nor); and that the AFTAC radionuclide surveillance plane has been sent and has found nothing related to the event." China may have conducted a subcritical experiment (hydrodynamic test), as the United States and Russia had done already several times since the CTBT was signed in 1996.

IMPROVED DETECTION AND IDENTIFICATION CAPABILITIES

Detection and identification improved greatly after the Kara Sea earthquakes near Novaya Zemlya of 1986 and 1997 and especially after the earthquakes in Tibet and Central Asia in the late 1960s. The seismic events in Eurasia that the United States claimed as problems or anomalies in 1972 are shown as a bar at the upper right of figure 12.6. The bar at the upper left of figure 12.6 shows the improved capability in 2009 to monitor seismic events on and near Novaya Zemlya.

In 1985 Ray Kidder published the relative frequency of U.S. underground nuclear explosions in Nevada from 1980 to 1984, shown in the

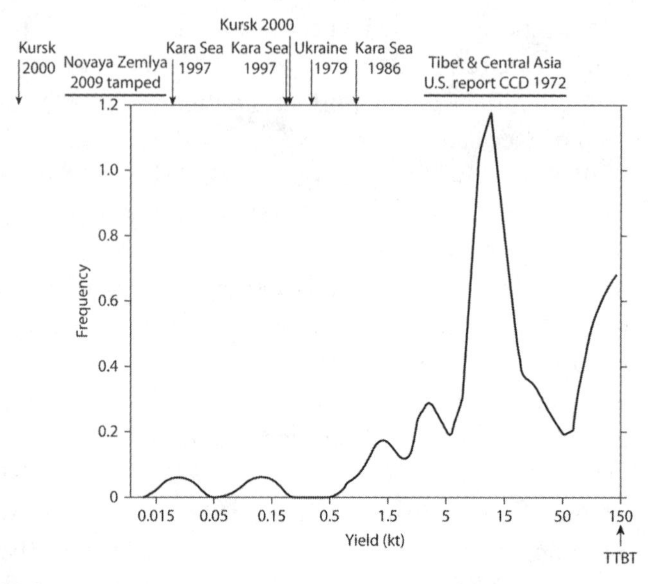

FIGURE 12.6

Problem seismic events are indicated at the top for various dates. Their yield in kilotons (kt), if they had been nuclear explosions, is shown on the horizontal axis at bottom. The bar indicates detection capability in 2009 for monitoring nuclear explosions at Novaya Zemlya. The lower part of figure, as modified from Kidder (1985), shows the frequency of U.S. underground nuclear explosions in Nevada from 1980 through 1984. TTBT indicates the 150-kiloton limit set by the Threshold Test Ban Treaty.

main body of Figure 12.6. He argued that peaks in his diagram near 7 to 20 kilotons and near 150 kilotons indicate tests with yields of high importance to the United States, whereas those with yields less than one kiloton were not. My work indicated the yields of past Soviet tests had a pronounced peak near 20 kilotons. The capability to monitor well-coupled explosions at Novaya Zemlya in 2009 is better than that of Kidder's smallest, less frequent explosions.

Conversely, nuclear explosions with seismic magnitudes of 4.5 to 5.5, like those of the so-called anomalous events in the CCD report of 1972, are near the peak of testing in figure 12.6. If they could not have been identified, the ability to monitor nuclear tests of military importance would have been poor in 1972. Even then, however, they were identified as earthquakes in our special study of 1976.

13

NEGOTIATING THE COMPREHENSIVE TEST BAN: GLOBAL MONITORING, 1993–2016

The Comprehensive Nuclear Test Ban Treaty (CTBT) was finally signed in September 1996 by Britain, China, France, Israel, Russia, the United States, and more than a hundred other countries. I paraphrase some of the following material from an extensive description by Ola Dahlman of Sweden and colleagues in 2009 about the negotiations that took place from 1993 to 1996.

NEGOTIATIONS

In August 1993, the United Nations' Committee on Disarmament (CD) issued a mandate to its Ad Hoc Committee on a Nuclear Test Ban to negotiate a comprehensive global ban on nuclear testing. Talks began in February 1994 with the establishment of a Working Group on Verification and another to deal with legal issues, the international organization, and entry into force of the treaty.

The new Working Group on Verification appointed Friends of the Chair, experts for specific verification issues. British geophysicist Peter Marshall, one Friend of the Chair, played a key role in 1995 and 1996 in formulating the global International Monitoring System (IMS). Peter, who died in 2012, deserves great credit for his long, intelligent, and honest efforts in nuclear verification going back to the late 1950s.

A close connection was established between the Working Group on Verification and the international Group of Scientific Experts (GSE), made up of government-appointed scientific experts, who had previously established three successive exchanges of scientific and technical information for the better detection and identification of seismic events.

All of the many countries represented in the 1995 negotiations supported a "zero yield" test ban, one in which no release of nuclear energy was permitted. In October 1995, presidents Clinton and Yeltsin announced that their two countries would work together to obtain a zero-yield treaty. Time was of the essence in the CTBT negotiations because many non-nuclear states were not likely to support an indefinite extension of the Nonproliferation Treaty (NPT) in 1995, when its twenty-five-year term expired, unless a CTBT was concluded in 1996. The original NPT specifically stated that the nuclear weapons states would work to achieve a CTBT. The United States also was anxious to see the NPT extended indefinitely, which happened in 1995. That same year the United States announced it would extend its moratorium on nuclear testing until the entry into force of a Comprehensive Nuclear Test Ban Treaty, provided the treaty was signed by September 30, 1996.

OPPOSITION TO A CTBT IN THE UNITED STATES

Republican Senator Jon Kyl of Arizona, an opponent of the CTBT, and Democratic Senator Harry Reid of Nevada introduced an amendment to the 1997 defense authorization bill that would have extended the president's power to order U.S. nuclear weapons tests beyond September 30, 1996. If passed, the amendment would have defied the treaty.

This divisive bill fortunately was defeated in June 1996 by a Senate vote of 53 to 45. Many people had thought that a Comprehensive Test Ban Treaty would be in place by then. An Associated Press report of June 26 stated, "Proponents of the amendment said they wanted to ensure the safe and dependable operation of the U.S. nuclear arsenal. Opponents said it could disrupt the Geneva talks [on the CTBT] days away from their conclusion." Had it passed, the amendment would have faced a certain veto by President Clinton.

For the presidential election of 1996 that pitted Robert Dole against Bill Clinton, the hawkish Republican platform on the CTBT stated: "To cope with the threat of the proliferation of weapons of mass destruction, the United States will have to deter the threat or use of weapons of mass destruction by rogue states. This in turn will require the continuing maintenance and development of nuclear weapons and their periodic testing. The Clinton Administration's proposed Comprehensive Test Ban Treaty (CTBT) is inconsistent with American security interests."

ACTIONS AT THE UN AND PROBLEMS WITH RATIFICATION OF THE CTBT

The chair of the UN's ad hoc CTBT negotiating committee delivered to the committee in May 1996 a draft treaty in which he attempted to capture consensus and bridge differences among the views of various nations. Some delegations refused to accept it as a basis for further negotiations, however, preferring to stay with the previous rolling document that contained many brackets with differing texts submitted by various countries. India, for instance, wanted a statement in the preamble about nuclear disarmament in a set time frame. This and other divisive issues became impediments to a CTBT because the work of the Committee on Disarmament and its Ad Hoc Committee on the CTBT required consensus by all parties for a treaty to go forward. This was and continues to be a great weakness of the UN's Committee on Disarmament.

While many countries, including the Russian Federation and the United States, stated in July 1996 that they were prepared to accept the draft treaty, India and Iran were not. Because a complete consensus could not be reached, the final text was presented in August 1996 merely as a report to the Committee on Disarmament. Even though the vast majority of states involved in the negotiations wanted to move forward, the text of the treaty itself was held hostage.

In an effort to bypass this holdup, Australia, a strong supporter of the treaty, introduced the same text to the UN General Assembly as a draft resolution on September 10, 1996. Decisions in the General Assembly do not require consensus and cannot be vetoed. Bhutan, India, and Libya voted against it, five countries abstained, and 158 voted in favor. Thus, on September 24, 1996, the CTBT was opened for signature. President

Clinton signed on behalf of the United States, as did representatives of Britain, China, France, and the Russian Republic.

In the United States, ratification of a treaty requires the Senate to give a two-thirds affirmative vote, called its "advice and consent." The U.S. House of Representatives is not involved directly in the ratification process. Following an affirmative vote by the Senate, the president must sign a treaty and submit it to the appropriate international body.

The treaty required that all forty-four states that possessed either nuclear weapons or reactors ratify the treaty before it could enter into force. As of early 2017, 183 countries had ratified the treaty, including France, the Russian Republic, and the United Kingdom, all three acknowledged nuclear weapons states. Included in the forty-four countries are China, Egypt, India, Iran, Israel, North Korea, Pakistan, and the United States, which have not ratified it as of early 2017. Although the United States and China signed the treaty, they still have not ratified it. India, North Korea, and Pakistan, which became nuclear weapons states after 1996, have neither signed nor ratified it.

INTERNATIONAL MONITORING

A separate international body, the Comprehensive Nuclear Test Ban Treaty Organization (CTBTO), was established along with a Technical Secretariat under the 1996 treaty. Formally, they are called provisional until the treaty enters into force. The treaty and its protocol describe terms under which on-site inspections of suspicious events could be proposed and undertaken once it enters into force.

To monitor compliance, an International Monitoring System (IMS) and an International Data Center (IDC), the most elaborate international verification system ever created, were set up within the CTBTO. A Global Telecommunication System transmits data in real or near real time to an international center in Vienna, Austria, that are available to member states. The major players—the United States, Russia, China, and France—receive large amounts of raw data from the center.

The International Monitoring System includes technologies that had been established by 1963 for monitoring tests in the atmosphere, oceans, and outer space. This ensured that, in addition to explosions underground, tests in all of these environments were covered after the treaty was signed in September 1996.

The IMS includes stations for monitoring seismic, hydroacoustic, and infrasound waves as well as measuring radioactive particles and noble gases produced by nuclear explosions. Hydroacoustic waves propagate very efficiently in the oceans; infrasound waves travel through the atmosphere at frequencies lower than those humans can detect. Noble gases are elements on the right-hand side of the periodic table. Because they do not react with other elements, they are difficult to contain, even for underground nuclear explosions, and therefore are easiest to detect.

A variety of countries now have certified laboratories that measure radionuclide particles and gases. Bomb-produced xenon and argon, two noble gases, can be detected at large distances in minute quantities (several atoms) from nuclear explosions in the atmosphere and from some underground nuclear tests. During the past twenty years, there has been a major advance in detecting minute quantities of those two gases.

Xenon isotopes were detected from the 2006 and 2013 underground explosions set off by North Korea. Those two events, as well as the 2009 nuclear test, all detonated under one mountain, were recorded by many seismograph stations and located very quickly by the International Monitoring Center in Vienna and the U.S. Geological Survey.

I add an historical footnote here about the leakage of xenon and other gases from past underground tests and the implications for monitoring the CTBT today. More than twenty years ago, Russia and the United States stated that a number of Soviet underground nuclear explosions at Novaya Zemlya generated noble gases that escaped and crossed international boundaries.

The United States and Russia interpreted leakage of gases differently under the terms of the 1963 Limited Test Ban Treaty. Nevertheless, the treaty states that both the English and Russian versions are equally valid. The differing interpretations hinge on the English phase "radioactive debris" and the Russian phrase "радиоактивных осадков," or "radioactive precipitates."

U.S. officials claimed that debris referred to both precipitates (fallout) and bomb-produced gases. It accused the Soviets of cheating on the LTBT when gases such as xenon were detected beyond the borders of the USSR. The Soviet Union claimed that the 1963 treaty did not cover noble gases because they were not precipitates. In backing up their claim, they published information on the leakage of gases from a number of underground nuclear tests at Novaya Zemlya, which they might not have done otherwise.

In the context of the 1996 CTBT, these findings indicate that preventing the escape of noble gases and avoiding their detection is much more difficult to accomplish than detecting the fallout of radioactive solids. Radioactive gases can be released either through drilling back into the site of an explosion at a later date or by purging the air in tunnels that had been used for a nuclear test. This information is a plus for better detection of clandestine testing, even for explosions conducted underground.

Another CTBT technology is the detection of infrasound, very low frequency sound waves in the atmosphere generated by above ground nuclear tests. It received new emphasis in 1996 after a long hiatus. William Donn, Frank Press, and Maurice Ewing had pioneered this technology at Lamont in the 1950s. Sensitive instruments developed to detect infrasound recorded many large U.S. and Soviet hydrogen bombs detonated in the atmosphere. Stations in South Korea using this technology recorded the much smaller North Korean tests of 2009, 2013, and 2016, even though they were conducted underground.

Another method uses hydroacoustic waves, which are propagated in the oceans. They were originally studied and monitored for the detection of submarines during World War II and the Cold War. Similar to sound waves that travel in the atmosphere, hydroacoustic waves propagate much more efficiently in the oceans than seismic waves in the solid Earth. France has been a leader in hydroacoustic and infrasound studies since the CTBT was signed in 1996.

The past twenty years have seen substantial improvements in monitoring underground testing. Commercial satellite imagery is openly available now with a resolution of better than three feet (one meter). A relatively new technology called INSAR (Interferometric Synthetic Aperture Radar) measures displacements of the Earth's surface such as those caused by earthquakes and underground explosions. Under the treaty, countries with one or more of these and other technologies, so-called national technical means (NTM), are permitted to use them to verify compliance.

They can present those data in requesting an on-site inspection of a suspicious event or a possible nuclear test.

The CTBT Organization interacts with what the treaty calls "member states" through designated national organizations and data centers. Experts from the United States participate in international working groups that set policy, priorities, and technical standards for the International Monitoring System and its data center. The Air Force Technical Applications Center (AFTAC) is the lead data center for the CTBT in the United States. The U.S. National Authority—the lead federal agency for the CTBT—is yet to be designated as its resolution has been stymied by bureaucratic infighting.

Some of the following material is taken from *The Comprehensive Nuclear Test Ban Treaty: Technical Issues for the United States* (2012), which I helped to write.

MISCONCEPTIONS ABOUT THE ROLE OF THE INTERNATIONAL ORGANIZATION

I would like to clarify two misconceptions I have encountered regarding the CTBT Organization—first, that it is responsible for *identifying* seismic events as nuclear explosions, earthquakes, or chemical explosions, and second, that it should deal with *evasive testing* such as decoupled explosions in large underground cavities.

Neither of these is correct. The treaty specifies that the last stages in monitoring—event identification and the possible attribution of a nuclear explosion to a particular country—are the responsibilities of national authorities, not the international organization. The United States insisted on this stipulation in the treaty negotiations. The reason for it is that attribution is a political act—one country asserting or implying that a treaty violation has occurred in another. The United States and other countries do not rely on the CTBT Organization to decide if a particular event was a nuclear explosion. AFTAC's classified capabilities are better than those of the CTBT Organization. U.S. nuclear monitoring—missions, capabilities, response times for analysis, and countries of concern—are different from those of the international organization.

The treaty and the operational manuals of the IMS specify the numbers and locations of stations, response times, and data quality for its international stations. These stations are distributed globally, without focusing on particular countries, although most of them had to be sited on land. Because existing test sites and countries of concern are located in the northern hemisphere where there is more landmass, the International Monitoring System has better coverage for those areas than, say, for southernmost South America. United States policy makers need to realize that AFTAC and other U.S. capabilities can focus both on better monitoring of regions of special concern to the United States and on global coverage.

OPERATION OF THE INTERNATIONAL MONITORING SYSTEM

Most stations of the International Monitoring System are operating now and are certified for their data quality and integrity (including tampering and data authenticity). The number of IMS stations of all five types grew from a few in 2000 to 83 percent of the full network of 321 stations in 2012 (figure 13.1). The full 100 percent of proposed installations has not been achieved because a few individual countries, such as India, have forbidden deployment of them on their territories.

The seismic network when completed will consist of fifty primary stations and 120 auxiliary stations (figure 13.1). Many of the primary stations are seismic arrays, which, unlike single stations, have the capability to determine the direction (azimuth) from which a seismic wave arrives as well as the distance to its source. Many arrays are very good at detecting small events and the seismic waves that directly follow the P wave such as pP, allowing an event's depth to be determined.

A number of the stations of the IMS are located in places that previously were not accessible to the United States. Several of them help to monitor a broad swath of countries of concern to the United States, stretching across Russia, China, the Middle East, and southern Asia. Several of the IMS seismic stations and arrays, such as that in Niger in west-central Africa, are among the most sensitive in the world. The very quiet

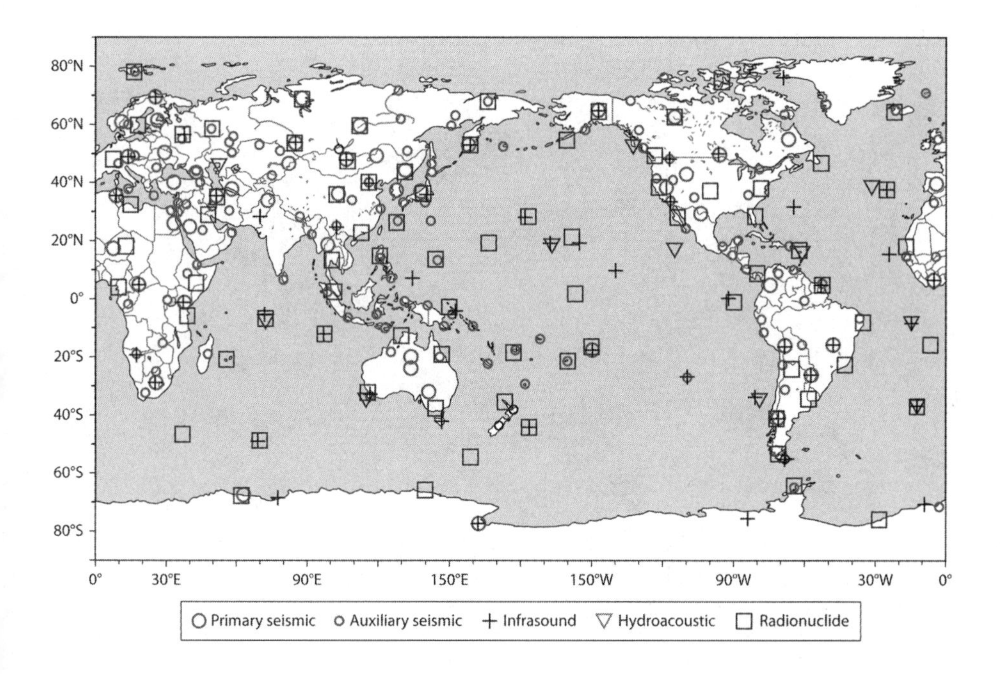

FIGURE 13.1

Five types of observation stations of the International Monitoring System.

Source: W.-Y. Kim, personal communication.

Niger array is located on old crust and the uppermost mantle of the Earth through which seismic waves propagate very efficiently.

One of the main products of the International Data Center is a Reviewed Event Bulletin (REB) for seismic events that is distributed to member countries—those that have signed the treaty—about every ten days. Significant events, like those at the North Korean test site, are distributed within hours.

Figure 13.2 shows the monitoring capabilities of only the primary IMS seismic stations that were operating as of late 2007. In it the globe is contoured for very high likelihood—that is, very high probability—of seismic events of various magnitudes being detected. Countries like the United States that want high confidence in identifying clandestine nuclear testing may pick a very high capability of detection, say 90 percent, as in figure 13.2. A country seeking to test but not wanting to be detected would

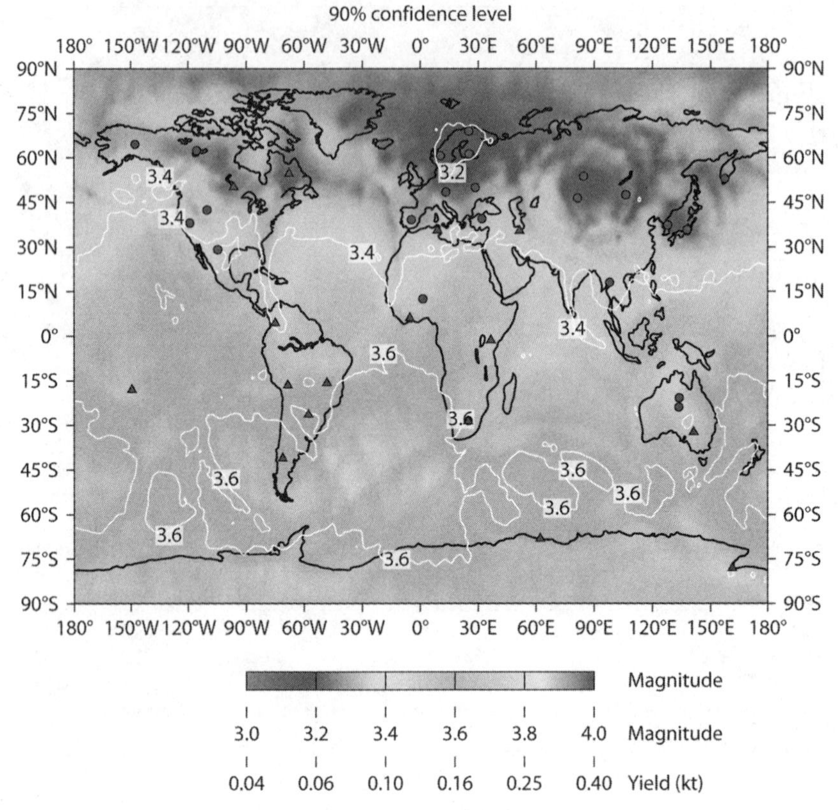

FIGURE 13.2

Detection capability in late 2007 of thirty-eight operating primary seismic stations of the International Monitoring System. Contours indicate the magnitude of the smallest event that would be detected with high likelihood (90 percent probability).

Prepared by Kværna and Ringdal of NORSAR with yields added for National Academies Report of 2012.

not chose a 90 percent capability because of the high likelihood of being caught, but would select a smaller number, say only 10 percent or 20 percent probability of being observed.

The contours of magnitude in figure 13.2 indicate a very good global and regional detection capability, better than that projected in the 1990s and early 2000s using several computer simulations. Enough stations are now operating that the figure is now based on actual P-wave readings.

Detection capabilities for seismic magnitudes and yields are better for countries of obvious concern to the United States than for many areas of the southern hemisphere. This is because more stations are located in the northern hemisphere than in the extensive southern oceans.

Capabilities shown in figure 13.2 are best at magnitude 3.2 for Scandinavia and 3.4 or better for most of Asia, Europe, North America, and North Africa. Those areas include states of obvious concern to the United States and all underground test sites except that of France in the South Pacific, which is now closed. Capabilities are somewhat poorer at magnitudes 3.6 to 3.7 for most of the southern hemisphere. Hydroacoustic waves, which propagate very efficiently to large distances, are much more useful for detection in most areas of the equatorial and southern oceans.

Figure 13.2 also indicates that capabilities for detecting yields of underground nuclear explosions are about 0.1 kilotons (100 tons) for regions with good seismic wave propagation like North Korea and most of Russia and China. Those yield capabilities are about three times worse for regions of poor seismic wave propagation like Nevada, eastern Turkey, and parts of Iran. While the capabilities for determining magnitude and yield for various areas of the world may be hard to take in all at once, they are all exceedingly good. The known yields of nuclear weapons in the U.S. and Russian strategic arsenals that are carried by intercontinental delivery systems are about a thousand times larger. Those capabilities are about a hundred times better than the yield needed to test the trigger for a fusion weapon (hydrogen bomb).

These capabilities do not include the use of additional data from either auxiliary IMS seismic stations or the huge number of high-quality stations of the International Federation of Digital Seismograph Networks. Digital recordings from many of those stations are now available in near real time over the Internet. Data are also available from large seismic networks in Canada, Japan, Taiwan, Turkey, the United States, and several European countries. About the same number of P-wave readings for the 2009 and 2013 North Korean nuclear tests were utilized for detection in near real time from those stations as from those of the International Monitoring System. Hence, they are a valuable supplement to international monitoring.

These seismic stations are used as well for other purposes, such as earthquake and tsunami warnings and studies of the Earth's crust and deep interior. Data from the seismic stations of the IMS also were used to study several great earthquakes such as those off Sumatra in 2004 and Japan in 2011. The radionuclide stations of the IMS have been (and continue to be) valuable in tracking radioactive leakage from the Fukushima, Japan, disaster of 2011.

To summarize, substantial improvements made in the past twenty years in U.S. and international capabilities to monitor underground nuclear testing include the following:

1. High-frequency seismic data from new stations at regional distances (up to 1000 miles or 1600 km) have been tested and implemented for the detection, location, and identification of seismic events, particularly those smaller than magnitude 4.
2. More broadband, high-quality seismic stations and arrays are now transmitting much greater volumes of digital data to data centers in near real time.
3. Major increases in computer power and data storage have led to the use of the entire waveforms of many past seismic events for determining better location and identification of events.
4. India, Pakistan, and North Korea, none of which has signed the treaty, are the only countries that have tested nuclear devices or weapons since the Comprehensive Nuclear Test Ban Treaty was signed in 1996. Tests as small as a fraction of a kiloton can be detected and identified at the Indian and North Korean test sites, which are situated in regions of very good seismic wave propagation. The 2006 North Korean test, which had a yield somewhat less than one kiloton, may well have been a fizzle with a smaller yield than expected.
5. Instrumentation to measure exceedingly tiny amounts of bomb-produced xenon has improved tremendously. The IMS network for monitoring xenon and other noble gases has gone from nearly nonexistent in 2000 to one that provides global coverage today. Several countries, including the United States, make additional measurements of noble gases.

Figure 13.3 shows the great improvements since 1990 in seismic detection of small explosions and earthquakes. The global threshold for detection has improved from about magnitude 4.5 in 1990 to 3.7 in 2007—as small as 0.25 kiloton for regions of good seismic wave propagation and about

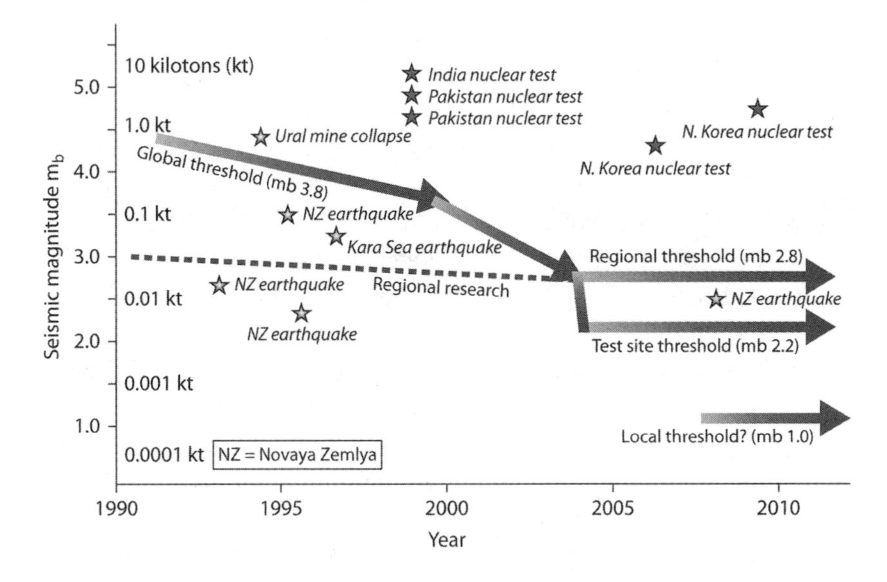

FIGURE 13.3

Improvement in seismic monitoring from 1990 to 2016. Since the scales are logarithmic, one unit of magnitude represents about a factor-of-ten improvement in seismic amplitude and yield. Yields assume full coupling in regions with good seismic wave propagation.

Source: National Academies Report, 2012.

0.75 kiloton for areas of poorer propagation. When hydroacoustic capabilities are included for the southern and equatorial oceans, the combined detection limits are better.

Detection thresholds for several regions of particular monitoring interest to the United States are now as good as magnitude 2.8. Capabilities to monitor the test sites of Russia at Novaya Zemlya and China at Lop Nor are better.

U.S. ATTEMPT TO RESTRICT DATA FROM INTERNATIONAL CENTER

The Air Force Technical Applications Center (AFTAC) is the U.S. National Data Center (NDC) for receiving seismic and other information from the International Data Center (IDC) for the CTBT. AFTAC uses the data internally and furnishes it to Department of Defense (DoD) contractors, other governmental agencies, and the weapons labs upon request.

Ralph Alewine of DARPA, who believed that data from the IDC should be restricted and not available in near real time to others, volunteered in May 1997 to draft an interim policy on the release of data in the United States. In June 1998, a White House interagency meeting chaired by Robert Bell of the U.S. National Security Council included a discussion on the release of data from the IDC. Despite skepticism from the National Security Council, Department of Defense officials maintained their position against the release of all data from the IDC, citing concerns about university scientists' using the data to make independent assessments of compliance issues and, in the event of their reaching a conflicting opinion, undermining the U.S. government's ability to charge violations or call for on-site inspections.

Interestingly, Alewine used the August 16, 1997, earthquake in the Kara Sea near Novaya Zemlya as an example of his point. It, of course, was the seismic event that was incorrectly identified by the U.S. government and remained so for more than two months. As discussed earlier, several university scientists, including me, identified it as a small earthquake, as did British seismologists, who stated shortly after the seismic event that it was a small earthquake. A number of scientists who studied the event at the U.S. weapons labs also agreed that it was not a nuclear explosion, but they were not allowed to comment publicly about the event for months.

This is a perfect example of why the Department of Defense and the U.S. intelligence agencies cannot be allowed to "bottle up" scientific data such as those from the IMS. Incorrect conclusions made by them have led to false claims about suspicious events of importance to U.S. national security and international relations. Such claims can have very dire consequences.

Fortunately, U.S. scientists can obtain an abundance of openly available seismic data for an event of questionable origin. Because the data are not classified, scientists from countries all over the world obtain data from the International Data Center through their own national authorities.

Although the interagency process normally is one of consensus, Robert Bell at the National Security Council evidently was frustrated that no progress had been made on the issue of data availability. In an extremely unusual move, he called for a vote of persons representing government agencies. Alewine's proposition was overwhelmingly defeated. Last-minute attempts by DoD to include a delay in the release of data and other restrictions were defeated as well. Hence, it became the position of the United States government in 1998 that all data from the IMS were to be openly available without restriction.

Although a few countries objected to the prompt release of radiological and some other data collected by the IMS, it is now possible for scientists like me in the United States to send a request directly to the IDC in Vienna, open an account, and obtain seismic data. I did just that in 2009 for seismic events near various test sites, as described in the next chapter.

14

MONITORING NUCLEAR TESTS SITES AND COUNTRIES OF SPECIAL CONCERN TO THE UNITED STATES

The Comprehensive Test Ban Treaty Organization (CTBTO) held a conference in Vienna in 2009 that focused on progress made in monitoring the treaty. Six months earlier it had solicited contributions from scientists utilizing data that the organization had collected and analyzed. I made two presentations at the conference using seismic data I received in response to their solicitation.

Meredith Nettles of Lamont and I obtained data from 2000 through 2008 that the CTBTO's monitoring arm had collected based on their seismic locations within 62 miles (100 km) of six sites used previously for nuclear testing. Identifying events at those sites is of great importance to policy makers. We found that all of those events could be identified as either earthquakes or explosions down to very low magnitudes in China, India, Pakistan, North Korea, and various other countries that are or may be capable of nuclear testing in the future.

Nettles and I examined thirty-eight seismic events of magnitude 3.3 and larger; the International Data Center of the CTBTO usually does not report events that are smaller than this. Most occurred near the former test sites of China, the United States, and Pakistan. No events were reported by the center near the Russian site at Novaya Zemlya or India's test site, both very quiet locations for earthquakes. The identification of small seismic events on or near Novaya Zemlya is described in chapter 12 and is not repeated here. All of the events we studied at the Nevada Test Site (NTS) were earthquakes. We identified all of those in North Korea, including two nuclear explosions and one earthquake.

Identification at magnitude $m_b > 3.3$ for five of the six test sites corresponds to a yield threshold of a small fraction of a kiloton if no serious attempts have been made to evade detection. The identification limit for Pakistan corresponds to about one kiloton, but it likely can be improved by examining high-frequency seismic waves, which our group has not done thus far.

CHINA'S TEST SITE AT LOP NOR

China conducted all of its nuclear explosions at its Lop Nor test site in the northwestern part of the country. Nettles and I found that the Lop Nor site had the greatest number of seismic events within 62 miles (100 km) of it, more than the other five test sites combined. Most of the seismic events that we studied near that site occurred at depths greater than 10 miles (16 km), clearly indicating they were earthquakes. High-frequency seismic waves also identify them and the remainder of the events as earthquakes (figure 14.1).

Although authors of several papers noted that the seismic event of March 13, 2003, of magnitude 4.3 to 4.7 was difficult to identify—that it was an anomalous or "problem" event—we found that it could be identified positively as an earthquake by five different methods. The smallest seismic event of magnitude 3.4 that we identified near Lop Nor would correspond to a well-coupled explosion with a yield of about 0.09 kilotons (90 tons). Work by others had indicated an identification capability that was even better, about magnitude 2.5, with a yield of 10 tons (0.01 kilotons).

MONITORING NUCLEAR TESTING BY NORTH KOREA

North Korea is one of the few countries that have not signed the Comprehensive Nuclear Test Ban Treaty (CTBT). It withdrew from the Nonproliferation Treaty in 2003. In 2005 North Korea declared that it possessed nuclear weapons, testing underground in 2006, 2009, 2013, and 2016 at Punggye-ri, a remote site in the northeastern part of the country. Five North Korean tests were well recorded and occurred very close to one another.

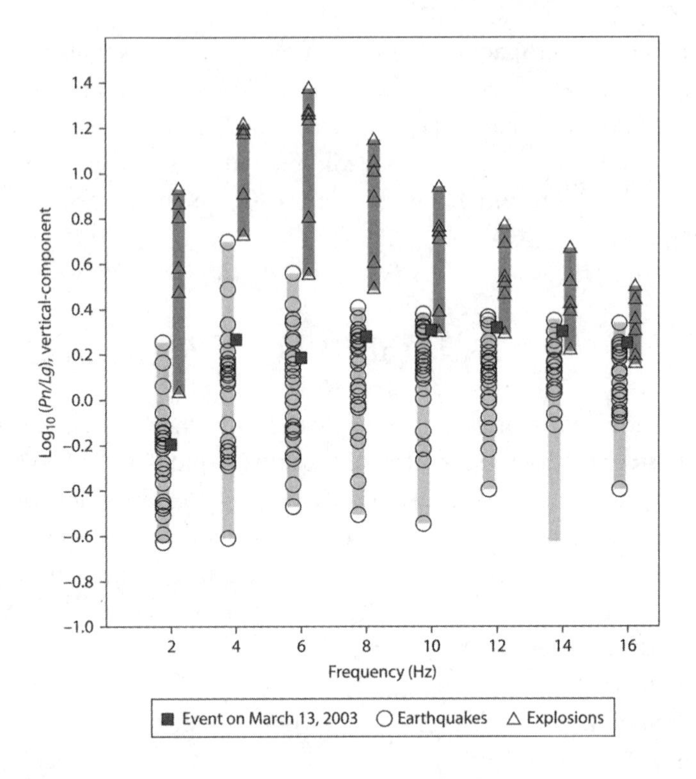

FIGURE 14.1

Measurements of high-frequency seismic waves for earthquakes and explosions near the Chinese test site at Lop Nor from 2000 to 2008. The log of the amplitude ratio of P to L_g seismic waves is shown on the vertical axis. Triangles denote earlier nuclear explosions at regional stations. Circles indicate earthquakes. Explosions have higher values on the vertical axis than earthquakes for frequencies of 4 and 8 Hz (cycles per second).

Source: Kim, Richards, and Sykes, 2009.

Although isolated, North Korea is a relatively small country and can be monitored readily by stations in South Korea, China, eastern Russia, Mongolia, and Japan. It consists largely of old rocks through which seismic waves are transmitted easily. Figure 14.2 shows the large numbers of stations of the International Monitoring Service that detected the North Korean nuclear explosion of 2013 with a yield of about 10 kilotons.

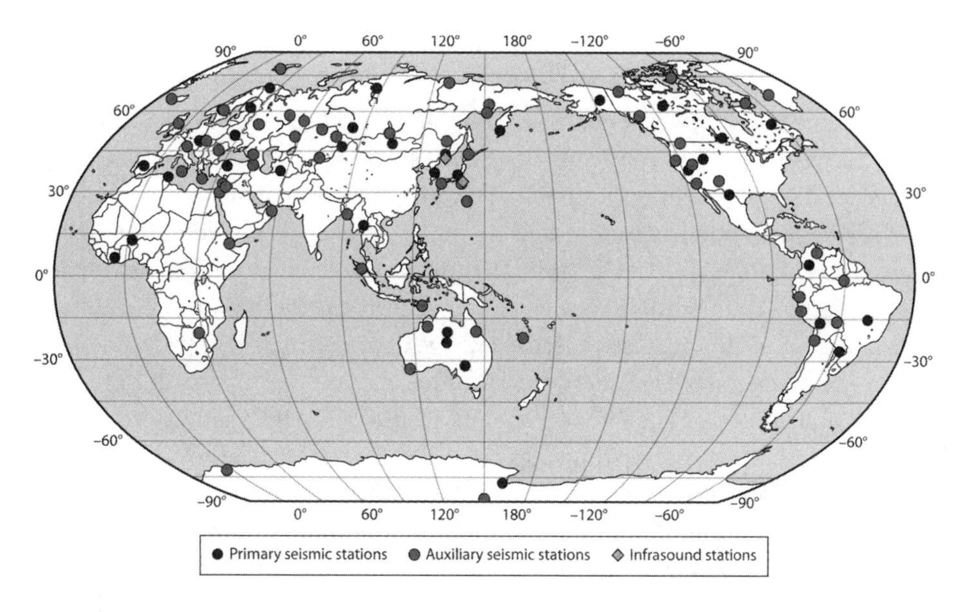

FIGURE 14.2

Seismic and infrasound stations of the International Monitoring System that detected the North Korean nuclear explosion of 2013.

Source: Comprehensive Test Ban Treaty Organization, 2013.

Kim and Richards calculated ratios of high-frequency seismic waves for events in North Korea, which are very similar to those shown for China in figure 14.1. Though not shown here, those ratios readily identify the five North Korean nuclear tests from 2006 to 2016 and a few rare, small earthquakes and chemical explosions. Seismograms of the North Korean tests are characterized by large P and a small Lg waves. Seismograms of small earthquakes exhibit the reverse: large Lg waves and relatively small P waves.

Well-coupled nuclear explosions in North Korea, like these five, can be detected and identified down to a few tens of tons—a small fraction of a kiloton. Stations in South Korea detected infrasound signals—low-frequency sound waves in the atmosphere—for the 2009 explosion. They also were detected in Japan and eastern Russia and South Korea in 2013. Bomb-produced noble gases were identified for the 2006 and 2013 explosions.

Monitoring a large underground explosion called the "chemical kiloton" at the Nevada Test Site in 1993 showed that two nonradioactive gases included in the explosive charge—sulfur hexafluoride and helium 3—could be detected in minute amounts along cracks and joints at the surface in the general vicinity of the shot point. Typically, they were detected at times of low atmospheric pressure and not at times of high pressure. Bomb-produced noble gases from an underground test are expected to behave similarly. Xenon isotopes were detected relatively late for the North Korean explosion of 2013, perhaps because leakage occurred many days later.

Some scientists have questioned whether the 2009 test by North Korea was, in fact, a large chemical explosion because radioactive gases were not detected in surrounding countries but were observed for the even smaller explosion of 2006. Huge chemical explosions detonated suddenly and of similar yield (several kilotons) and seismic magnitude to that of the 2009 test are exceedingly rare. Most are detonated sequentially over times as long as a second to break rock effectively and to reduce damage to nearby structures. In addition, transporting that much chemical explosives to the North Korean test site would have required huge numbers of trucks or train cars, which presumably would have been detected by satellite imagery. I doubt the 2009 test was anything other than a nuclear explosion.

North Korea announced all five tests ahead of time as nuclear and made no attempt to hide that fact, except possibly by detonating them at deeper than normal containment depths. They wanted other countries, particularly the United States, to know that they had nuclear devices. The main danger of North Korea's possessing nuclear weapons is not that they might be used to attack the United States or its allies, because the United States undoubtedly would counter that massively. Instead, the real danger is that materials, plans for weapons, or nuclear weapons themselves could be sold or transferred either to other countries or to terrorists. Countering those threats deserves higher priorities.

INDIA'S NUCLEAR TESTS

The main reason India conducted a nuclear explosion first in 1974 and again in 1998 was fear not of Pakistan, but mainly of China, which captured territory from India during the Himalayan war of October 1962.

China tested its first nuclear device in 1964 and its first multimegaton thermonuclear device in 1967.

The modern Pakistani station at Nilore provided near real time seismic data for the nearby Indian nuclear tests of May 11, 1998, to the IRIS seismological data center in the United States. The test site at Pokhran is located in northwestern India near the Pakistani border.

India and Pakistan, which have signed neither the CTBT nor the Noproliferation Treaty, do not send data to the International Monitoring Center. For India this likely stems from its not obtaining the language it wanted in the treaty in 1996, coupled with its unstated desire to test more sophisticated nuclear devices than the one it first detonated in 1974. The nationalist BJP party authorized the 1998 Indian tests when it came to power. Indian scientists later published high-quality seismograms for the Indian tests of May 11 and the Pakistani tests of 1998. Published reports indicate that the United States was caught off guard by the first Indian tests in 1998.

Routine transmission of data from seismograph stations in India would be very helpful in monitoring Pakistan, China, Iran, and other parts of southern Asia. Near real time transmission of data would have aided assessment of the very damaging giant earthquake and tsunami off Sumatra and the Andaman Islands in December 2004. The seismic array in Niger in central Africa is one of the best monitoring stations for explosions and earthquakes in southern Asia, including India and Pakistan.

The Indian prime minister said on May 11, 1998, that the yields of its tests earlier that day were as expected and that they consisted of fission, low-yield, and thermonuclear devices. He said they were contained and that no radioactivity was released into the atmosphere. This was followed six days later by a statement from Indian scientists that "there was no *harmful* radioactivity [my italics] from the contained nuclear tests."

The three nearly simultaneous nuclear explosions by India on May 11, 1998, produced craters and damage that can be seen on unclassified satellite images as well as in photographs released by India. It is hard to believe that radioactive materials, especially noble gases, were not vented, because explosions shallow enough to produce craters often leak radioactive gases. India's explosions at Pokhran in 1998 were within 2 miles (3 km) of the crater produced by its first nuclear test in 1974.

The magnitude of 5.3 determined from global data for the three simultaneous explosions on May 11, 1998, indicates a total yield of about 14 to 20 kilotons. The combined yield for the three tests as announced by the India's Bhabha Atomic Research Centre, however, was considerably higher, 58 kilotons. It is hard to comprehend that the yield based on global data could have been underestimated by a factor of three to four. India's official yield for the 1974 test, called the *Smiling Buddha*, was 20 kilotons, also higher than estimates made by other countries.

India obtained the plutonium for its 1974 explosion from a non-safeguarded reactor that followed a Canadian design. The United States provided the heavy water moderator for the reactor. India claimed the 1974 test was a peaceful nuclear explosion and that it did not violate the Limited Test Ban Treaty, which it had signed. Still at issue is whether India violated its peaceful use agreements with Canada and the United States.

Debate has raged in the media among Indian nuclear scientists about whether their country obtained a dependable thermonuclear (hydrogen) device through its tests in 1998. Its relatively small seismic magnitude indicates it may have failed to produce its full yield. Thermonuclear devices typically are triggered by fission explosions with yields of about 10 to 20 kilotons. Some Indian nuclear scientists, in fact, claim India needs to test a larger thermonuclear device to make sure they have an H-bomb capability. Others in India say they do not need such a test. India, which has not tested a nuclear weapon since 1998, might well have tested a boosted fission weapon at that time.

India and Pakistan each have aircraft that can carry nuclear weapons to the other. India has tested ballistic missiles that can reach China and, of course, Pakistan. Future nuclear explosions by either India or Pakistan could lead the other to test. China then might test as well.

The use of nuclear weapons by India, China, or Pakistan against one another represents a great nuclear threat today. Huge populations are at risk. Each continues to pursue more advanced delivery systems. Pakistan's nuclear arsenal continues to grow. Of particular concern is that neither India nor Pakistan may have mechanisms on their nuclear weapons to insure against unauthorized use or unplanned nuclear detonation during an accident or fire.

India also claimed soon afterward that it tested two small experimental nuclear devices with yields of 0.2 and 0.6 kilotons nearly simultaneously two days later, on May 13, 1998, at 06 hour 51 minutes. I know of no seismic stations that recorded those claimed very small explosions. The Nilore Pakistani station had a signal-to-noise ratio of about 1000 for the combined explosions two days earlier on May 11. My colleague Paul Richards found that Nilore had the capability to detect a combined yield on May 13 of about 0.025 kilotons (25 tons). Those estimates of yield assume they were conducted in the same rock type as those on May 11. This is much smaller than the announced Indian combined yield of 800 tons. India described one of the tests on May 13 as having been detonated in a sand dune, a poor coupling material. If so, the combined yield could have been one to a few hundred tons, still smaller than the combined announced yields.

Possible explanations of these inconsistencies are: (1) the two events detonated but were much smaller than one to a few hundred tons; (2) either they both failed to detonate or only the one of 200 tons did; (3) the yields were poorly calibrated by India, as they were for the explosions of May 11; or (4) the yields were exaggerated.

The implications for monitoring are that if craters were formed today as claimed by India for the two very small explosions on May 13, 1998, they would be detectable by satellite imagery and synthetic aperture radar (INSAR). If radioactivity was released, such an event set off today should be detectable. In any case, if one or both of those explosions on May 13 occurred in a sand dune, as claimed by India, it would be difficult to hide because dry sand is not a good medium for hiding a sub-kiloton clandestine nuclear explosion at shallow depth.

PAKISTAN'S NUCLEAR PROGRAM AND TESTS

The loss of East Pakistan (now Bangladesh) in a civil war may have triggered a 1972 political decision in Pakistan to begin a secret nuclear weapons program. India's 1974 nuclear explosion likely gave additional urgency to the program. Pakistan and India clashed several times after the partitioning of the two countries in 1947 when each gained independence from Britain. The two countries have fought over territory in Kashmir many times. Pakistan's conventional forces being inferior to those of India, whose population is considerably greater than Pakistan's, likely contributed as well to its quest for nuclear weapons.

Pakistan announced it conducted two series of nuclear explosions in late May 1998, soon after India tested on May 11. It said that the first set consisted of five explosions, which were detonated nearly simultaneously. Many people have speculated that Pakistan wanted to "up India's claim" of five tests earlier in May. Pakistan's two sets of explosions were well recorded internationally. The combined yield of the first set was about 10 kilotons, the second about 5 kilotons.

Pakistan gained uranium enrichment technology from many sources. It also obtained knowledge about nuclear weapons and missile technology from China. A. Q. Khan, widely regarded as the father of Pakistan's nuclear program, learned about centrifuge enrichment when he worked at the Uranium Enrichment Corporation, URENCO, in the Netherlands. URENCO provides enriched uranium for many European nuclear power plants. Khan stated that Pakistan began uranium enrichment in 1978 and produced highly enriched uranium by 1983. He reportedly stated in an interview in a speech in January 2010 that Pakistan "had become a nuclear power" in 1984 or 1985.

In 1990 President George H. W. Bush failed to certify that Pakistan did not possess a nuclear explosive device. The United States hesitated on several occasions to declare that Pakistan had a nuclear weapons program because of its desire to obtain Pakistani help in combating the Russian invasion of Afghanistan and in pursuing a war against the Taliban and Al Qaeda.

Khan's nuclear assistance extended to several countries was likely the most dangerous act of proliferation thus far during the nuclear age. The U.S. Congressional Research Service reported in May 2012 that Khan and his network had sold centrifuge technology for uranium enrichment to Libya in 1984. Libya revealed and gave up its nuclear facilities in December 2003. It then signed the CTBT and the Nonproliferation Treaty. One reason Libya did so was fear that the United States, Britain, and other countries would invade it. Materials that had clearly originated from Khan's group in Pakistan were found in documents that Libya surrendered. Khan announced in 2005 that he and his associates had sold centrifuge technology to North Korea. They may have offered similar assistance to Iran, Egypt, and perhaps Syria and Saudi Arabia.

In their 2009 book *The Nuclear Express*, Thomas Reed and Danny Stillman say they believe Pakistan tested its first nuclear bomb in 1990 at China's Lop Nor site. China considered Pakistan to be a regional ally. Others have expressed more uncertainty about the amount of Chinese nuclear assistance.

Reed was a former nuclear weapons designer at Livermore, secretary of the Air Force, and a special assistant to President Reagan for national security policy. Stillman worked at Los Alamos for decades in nuclear design, diagnostics, and testing. He directed the Los Alamos Technical Intelligence Division for thirteen years. They were experts on Soviet and Chinese nuclear weapons. Reed and Stillman also describe how various countries acquired nuclear expertise from others who already possessed nuclear weapons.

In an interview in *U.S. News and World Report*, Reed said, "There are numerous reasons why we believe this to be true, including the design of the weapon and information gathered from discussions with Chinese nuclear experts." Reed claims that the Pakistanis were so quick to respond to the Indian nuclear tests in May 1998 because the Chinese had already helped them prepare for a test to be conducted within Pakistan. Reed also noted, "It only took them two weeks and three days [to respond and test]." Reed and Stillman claim the Pakistani test on May 28, 1998, likely was a single explosion, not five. It was of an advanced HEU (highly enriched uranium) design. It is clear from seismic data that Pakistan conducted tests at two separate sites in May 1998.

The United States was concerned several times about Pakistan's ongoing production of nuclear weapons and its political instability that could lead to weapons' falling into the hands of insurgents and perhaps terrorists. Pakistani officials and high military officers have stated many times, however, that their nuclear weapons are secure.

Nettles and I worked on the identification of seismic events located in the vicinity of the two Pakistani nuclear tests. Pakistan is second only to Lop Nor in the number of earthquakes per year within 62 miles (100 km) of their test sites. We obtained focal mechanism solutions of the very long period type for ten moderate-size seismic events from 1980 to 2008, which indicated that they were earthquakes, not explosions. We found that all but one of the Pakistani earthquakes occurred at depths of 12 to 44 miles (18 to 61 km), indicating that they must have been earthquakes. Several other seismic events in Pakistan were identified as earthquakes using the Ms-m_b technique.

Identification of small earthquakes in Pakistan could be much improved by analyzing high-frequency seismic waves, which the Lamont group has not done. Seismic data from nearby stations in Afghanistan, Oman, and the United Arab Emirates have become available recently. Data from Indian and Iranian stations would be valuable as well to better monitor Pakistan. It is understandable that Pakistan did not make seismic data available for its tests later in May 1998. Monitoring, of course, needs to assume that a country testing a nuclear device will not provide seismic data from its own stations, at least not immediately.

ISRAEL'S NUCLEAR PROGRAM

Although none of the countries in the Middle East has declared that it possesses nuclear weapons, Israel maintains a policy of nuclear ambiguity. In the arms control community, it is very widely thought that Israel started to acquire nuclear weapons in the 1960s and now possesses about eighty. Its Dimona reactor, constructed with French help, went critical in December 1963, according to Reed and Stillman.

Israel, France, and Britain went to war with Egypt in October 1956 to reopen the Suez Canal, which had been seized by President Gamal Abdel Nasser. President Eisenhower publicly opposed the operation, and the Soviet Union issued an ultimatum to Britain, France, and Israel to desist, threatening actions that would menace the existence of Israel. Humiliated by the two superpowers, the three countries believed they could no longer count on American support.

Reed and Stillman say, "It is quite clear that France and Israel undertook a joint nuclear weapons program in the aftermath of Suez. That relationship, on a commercial basis, continued for decades." The Dimona reactor in Israel, which produced plutonium, was very similar to the reactor at Marcoule, France. Reed and Stillman state, "Some wags have noted that on

February 13, 1960, the two nations went nuclear with one test [in Algeria]." If true, this would explain why Israel has a modern nuclear arsenal without having tested by itself in the 1960s. Israel, a small country, does not have a test site of its own and is easily monitored from surrounding countries.

ISRAELI AND SOUTH AFRICAN NUCLEAR COOPERATION

Reed and Stillman claim, "After the Yom Kippur War of 1973, Israel and South Africa established closer nuclear ties, and Israel was assured of a uranium supply well into the future." In 1997 South African officials announced that Israel had helped it develop nuclear weapons.

On September 22, 1979, a U.S. satellite detected a "double flash" typical of a small nuclear explosion over the southern oceans. U.S. hydrophones on Ascension Island in the equatorial Atlantic, designed to be used for listening for underwater sound, picked up signals that were consistent with a nuclear explosion in the Indian Ocean on or near South Africa's Prince Edward Islands. Seymour Hersh, an American investigative journalist, stated in 1991 that a flotilla of South African and Israeli military ships had been tracked by the U.S. National Security Agency to a site near the Islands. Some media reports, apparently based only on the satellite data, placed the event incorrectly in the South Atlantic.

Many people in the U.S. weapons labs and the Defense Department concluded then, and continue to think, that it was a small nuclear explosion, perhaps of a neutron bomb that Israel thought it needed for close combat with tanks and Arab forces as occurred during the 1973 war. The double flash could not have occurred at a less opportune time for the Carter administration as the president was about to submit the second Strategic Arms Limitation Treaty (SALT II) to the Senate and to run for reelection on his success with nonproliferation. At the time, American foreign policy was in jeopardy in Iran following the overthrow of the shah and the capture of American hostages.

Carter convened a panel of scientists, including Lamont's William Donn, to examine classified data related to a possible small nuclear explosion. Their mandate, however, was to examine only technical data. Hersh says, "Our capturing it [the double flash] fortuitously was an embarrassment, a big political problem, and there were a lot of people who wanted to obscure the event." The panel, which was chaired by Jack Ruina of MIT, concluded that the flashes probably were not caused by a nuclear explosion but perhaps by a micro-meteorite hitting the satellite. Others attributed it to a small nuclear explosion close to the ground by Israel in cooperation with South Africa. Nevertheless, the global monitoring system is certainly capable of detecting and identifying such an event today. That is one reason the International Monitoring System includes more than just seismic monitoring.

Israel signed but has not ratified the Comprehensive Nuclear Test Ban Treaty (CTBT), nor has it signed the Nonproliferation Treaty. Israel does furnish data to the International Monitoring Center in Vienna, including information from a seismic array.

SOUTH AFRICAN NUCLEAR PROGRAM

South Africa developed first-generation nuclear weapons during the apartheid era. In the 1970s it became concerned about armed forces that opposed its regime in the former Portuguese colonies of Angola and Mozambique, in Southern Rhodesia (now Zimbabwe), and in South-West Africa (now Namibia). South Africa was further distressed by the introduction of Cuban troops into Angola and by warfare in South-West Africa. These forces, as well as embargoes against it, led South Africa to pursue a nuclear weapons program.

Soviet satellites detected South Africa's development of a nuclear test site in the Kalahari Desert in July 1977. They and the United States put pressure on South Africa not to test a crude, first-generation uranium weapon, which it did not do.

Late in 1993 President F. W. de Klerk disclosed details about South Africa's abandoning its nuclear weapons program in 1989 and dismantling six gun-type uranium 235 nuclear weapons, similar to the one dropped on Hiroshima in 1945. In is unknown whether those crude weapons would have worked, but many people assumed that they would. The government of de Klerk did not want those weapons to fall into the hands of a new government under Nelson Mandela. The International Atomic Energy Agency declared that the South African shafts in the Kalahari Desert had not been used for nuclear testing and, in 1993, that they had been rendered useless for nuclear tests.

South Africa is the only country that has destroyed its inventory of nuclear weapons. Sweden actively planned to acquire nuclear weapons in 1945 but then halted those programs in 1968. On June 1, 1996, Ukraine became "nuclear free" after returning the last of its 1900 Soviet-era strategic nuclear weapons to the Russian Federation. All three countries went on to sign and ratify the Nonproliferation Treaty and the CTBT.

MONITORING THE NEVADA TEST SITE

Nettles and I also studied seismic events from 1992 to 2008 within 62 miles (100 km) of the Nevada Test Site (NTS). Focal mechanism solutions of the very long period type, measurements of Ms-m_b, and data on high-frequency waves at regional stations indicate that all seismic events near NTS of magnitude greater than 3.3 could be identified as earthquakes. Unlike earthquakes near the Chinese and Pakistani test sites, shocks on and near NTS are quite shallow and hard to identify using just the seismic waves pP and sP (figure 3.1). Nevertheless, the three other methods sufficed for positive identification of them as earthquakes even though NTS is a region characterized by poor propagation of seismic P waves.

IRAN AND THE MIDDLE EAST

The 2012 National Academies report *The Comprehensive Nuclear Test Ban Treaty* examined the monitoring of the CTBT for Iran and other parts of the Middle East. Understandably, those countries are of major concern to many in terms of their possibly acquiring and testing nuclear weapons. Iran signed but has not ratified the CTBT. About a decade ago, it allowed the International Monitoring Service (IMS) to operate seismic stations on its territory. After the IMS certified them, however, data were no longer transmitted abroad even though the stations still exist.

Major concern about Iranian nuclear intentions led me to devote particular attention recently to the geology and earthquakes of Iran. If Iran acquires nuclear weapons, Egypt, Saudi Arabia, Turkey, and some Gulf states may also decide to do so. Iran still has some centrifuges running that produce enriched uranium, claiming it enriches materials solely for the generation of nuclear power. Its reactors and the amounts of enriched uranium are subject to the agreement reached in 2015 by Iran, the five main nuclear powers, Germany, and the European Union.

Since earthquakes occur often in Iran, distinguishing their seismic signals from those of underground nuclear explosions is of concern to many nations. Hence, I now describe the tectonic settings of Iranian earthquakes pertinent to their identification.

Iran is largely a region of continental convergence caught between the Arabian and Eurasian plates. Similar deformation extends into northeastern Iraq, the Caucasus, eastern Turkey, and Turkmenistan. Compression causes the continental crust of Iran to be squeezed outward into its surrounding countries and the southern Caspian Sea. In contrast, the oceanic plate beneath the Arabian Sea is being subducted along the Makran plate boundary of southeastern Iran and southern Pakistan. Subduction is inhibited today elsewhere in Iran.

The Arabian plate underthrusts the southwestern side of the Zagros Mountains along the Persian Gulf in southwestern Iran and northeastern Iraq. The crystalline rocks of the crust beneath the Zagros are part of the Arabian plate. Oceanic crust was subducted along the northeastern side of the Zagros about 80 million years ago. The suture zone and the Main Reverse Fault that remain from that former subduction are largely devoid of earthquakes. A second period of crustal shortening occurred in the Zagros during the last few million years. GPS measurements indicate that active crustal shortening is concentrated along the frontal, southwestern part of the Zagros belt.

Iran has a long history of damaging and deadly earthquakes. In 1997 M. Berberian stated that earthquakes had killed 126,000 Iranians during the previous hundred years. In their 1982 book *A History of Persian Earthquakes*, N. N. Ambraseys and C. P. Melville list many shocks and the destruction of cities going back thousands of years. The Tabas earthquake (magnitude Mw 7.4) of September 1978 in east-central Iran and the Rudbar-Tarom earthquake (magnitude Mw 7.7) of June 1990 in northwestern Iran were the most catastrophic earthquakes in Iran during the twentieth century, killing more than 20,000 and 40,000 people, respectively. Financial losses from the June 1990 earthquake were about $7.2 billion, about 10 percent of Iran's gross national product. About 30,000 deaths occurred in the moderate-size Bam earthquake of December 2003 (magnitude 6.6) in southeastern Iran. Poor construction and the shallow depths of those earthquakes contributed to the large loss of life and high damage.

Large earthquakes (figure 14.3) do not occur randomly throughout Iran but are concentrated in the Alborz Mountains in the north, in northwestern Iran, along faults surrounding the otherwise nearly nonseismic Lut block between 27 and 34 degrees north and 57.5 to 60 degrees east in eastern Iran, and in the Kopet Dag Mountains along Iran's northeastern border with Turkmenistan. Since 1918 most of the earthquakes along the Zagros Mountains have not exceeded magnitude 6.5, but many shocks of small to moderate size have occurred there. Central Iran, which is located between these zones of higher activity, has few large or moderate-size earthquakes.

Other large earthquakes have occurred in adjacent countries. The largest, of magnitude 8, occurred in 1945 off the coast of Pakistan, not Iran, along the Makran subduction zone (figure 14.3). Uplifted terraces along the coast in the Iranian part of Makran are indicative of past great shocks.

Monitoring possible nuclear testing by Iran involves the identification of seismic signals of possible explosions as distinct from those of the many moderate-to-small earthquakes that occur every year. Accurate determination of depths is key to identifying many seismic events as earthquakes.

In 2004 M. Tatar and colleagues obtained the most accurate depths of seismic events in the Zagros Mountains using local portable stations. The earthquakes they studied occurred within the upper 5 to 10 miles (8 to 16 km) of the crust; none were deeper than 20 miles (32 km). They report that these earthquakes are likely located in the upper part of crystalline crust below the very thick sedimentary layers of the Zagros. The thick Hormuz salt, which is about 430 to 600 million years old, is found at the base of those sediments and the top of the crystalline basement.

This is fortunate for verification because most of the earthquakes they examined beneath the Zagros were much deeper than any nuclear explosions that could be detonated there. Hence, accurate determinations of

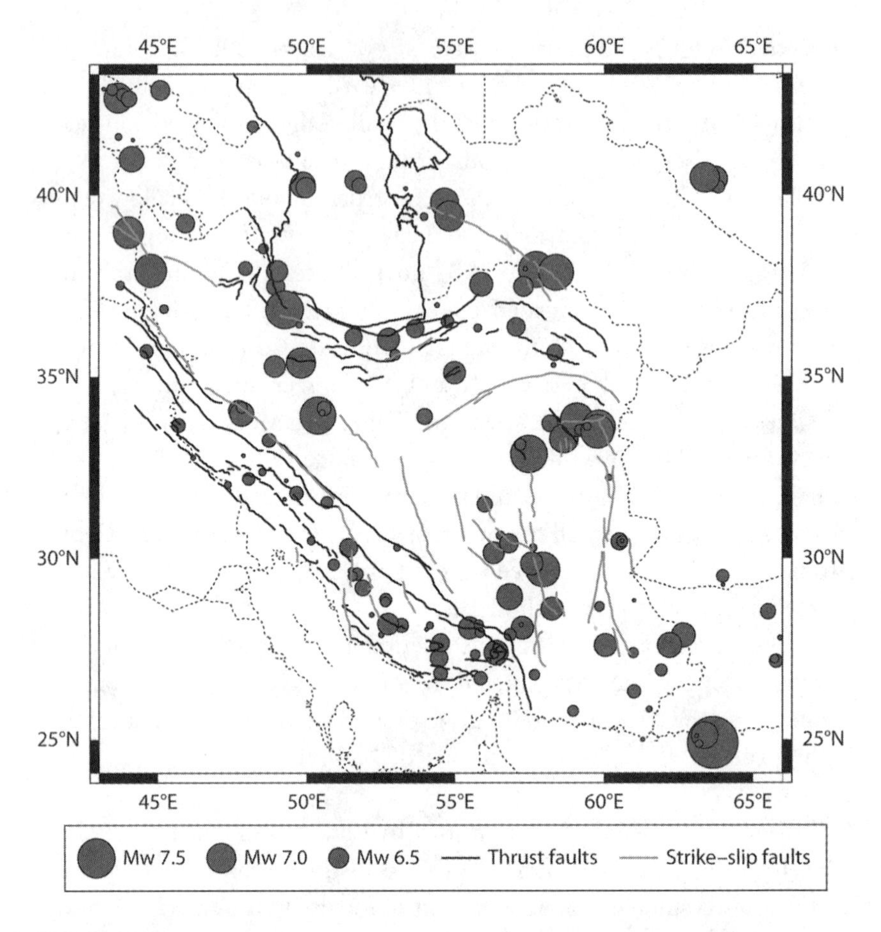

FIGURE 14.3

Locations and magnitudes of large earthquakes (Mw>5.4) in Iran and surrounding regions from 1900 to 2009.

Source: R. Engdahl, personal communication, 2013.

depths using data from local seismic stations are exceedingly valuable for the identification of events as either explosions or earthquakes.

Many sources of data on Iran's earthquakes and geology are accessible to those interested in seismic verification. The great loss of life and destruction from past shocks led the Iranian government many decades ago to work on reducing earthquake losses through the operation of seismic networks, engineering of buildings to better withstand strong shaking, geologic studies, and mapping of active faults. Many papers on these topics, as well as on Iran's petroleum resources and their geology, have been published over the past century.

In 2009 Michael Pasyanos and colleagues made an extensive study of thousands of paths that seismic waves travel from earthquakes in the Middle East to tens of regional seismic stations. They also used data from stations and earthquakes within Iran. In 2013 Mark Fisk of Alliant Techsystems and Scott Phillips of the Los Alamos Lab studied hundred of thousands of paths traversed by four different types of seismic waves in Asia, Europe, and the Middle East. The two studies show in detail how well or how poorly seismic waves are detected at various frequencies for many seismic paths and how well events can be identified as being either explosions or earthquakes. Both studies provide a rationale for choosing data from seismic stations best suited to examining and identifying future seismic events.

IRAQ

The United States invaded Iraq in 2003. Government claims that Iraq possessed an active program to develop nuclear weapons turned out to be false. Another false accusation was that Iraq had obtained uranium from Niger in central Africa. France, however, gets uranium from Niger, a former colony, and carefully controls Niger's export. The United States is said to have destroyed Iraq's uranium (yellow cake) earlier, during the first Gulf War. Another false claim was that Iraq had conducted one or more decoupled nuclear explosions beneath a large lake in 1989. Iraq subsequently signed and ratified the CTBT.

DEALING WITH PROBLEM SEISMIC EVENTS GLOBALLY: A SUMMARY

The numbers and sizes of various problem or anomalous seismic events decreased dramatically from 1960 to mid-2009 (figures 14.4 and 14.5).

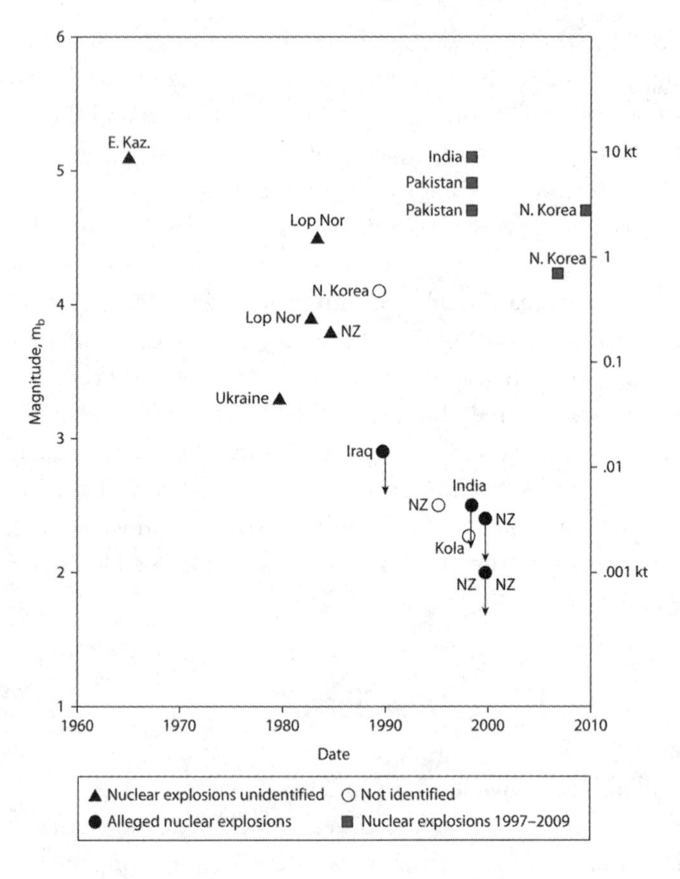

FIGURE 14.4

Unidentified nuclear explosions, alleged explosions, unidentified seismic events, and identified nuclear explosions by India, Pakistan, and North Korea from 1997 to mid-2009. Seismic magnitude is at the left side and yield in kilotons (kt) at the right. Downward-pointing arrows indicate that events were equal in size or smaller.

It should be remembered that these are a tiny fraction of the earthquakes and chemical explosions that are reported every year. Most problem seismic events are small, occurring near the lower end of seismic detectability at the time.

Figure 14.4 shows unidentified nuclear explosions, alleged nuclear explosions, and seismic events that were not positively identified initially, as well as recorded nuclear explosions by India, Pakistan, and North

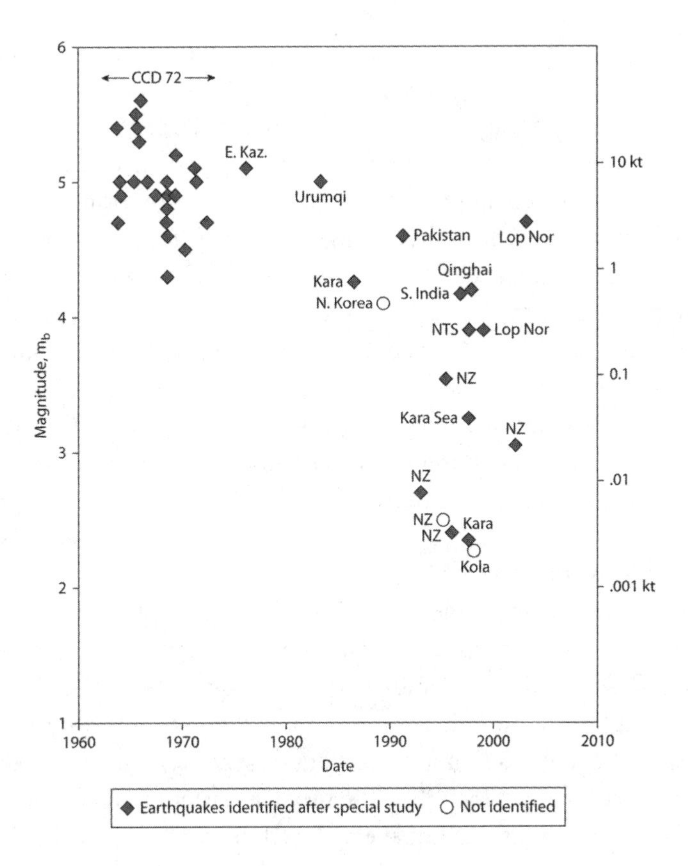

FIGURE 14.5

Events not identified and problem (anomalous) seismic events identified as earthquakes through special studies: Eastern Kazakhstan, E Kaz; Kara Sea, Kara; Kola Peninsula, Kola; Nevada Test Site, NTS; Novaya Zemlya, NZ. CCD 72 are seismic events from a U.S. document tabled in 1972 at the UN's Conference of the Committee on Disarmament.

Source: Sykes and Nettles, unpublished poster, 2009.

Korea from 1970 to mid-2009. Nuclear explosions that were not identified soon after they occurred are denoted by solid triangles. They decreased in size from about 8 kilotons in 1965 to about 0.3 kilotons in 1985. We found no unidentified nuclear explosions or any unidentified events larger than magnitude 2.5 (or 0.005 kilotons, 5 tons, if they were nuclear) after 1989.

Solid squares indicate nuclear explosions by India, Pakistan, and North Korea. They are the only countries that tested after the Comprehensive Nuclear Test Ban Treaty was open for signature in September 1996. The three did not sign the treaty. India and Pakistan last tested in 1998.

Seismic waves were not found, despite diligent searches, for one alleged nuclear explosion on September 19, 1989, in Iraq. That claim, which was publicized by dissidents, is almost certainly false. The seismic event in North Korea on May 22, 1989, which concerned some people in the United States, could not be identified at the time. There is no indication, however, that North Korea tested as early as 1989. Seismic waves were not detected for two suspect Russian tests on September 8 and 23, 1999, at Novaya Zemlya, as reported by Bill Gertz in the *Washington Times*. Noise measurements indicate that those events were not larger than magnitude 2.0 and 2.5. Both may have been permitted subcritical tests—that is, ones with no release of nuclear energy.

Figure 14.5 shows problem or "anomalous" seismic events that were later identified as earthquakes through special studies. Their magnitudes decreased from 5.6 in the mid-1960s to 2.7 by 2009. If they had been nuclear explosions, which they were not, their detectability in terms of yield improved over time from about 30 kilotons to 0.003 kilotons (3 tons). Nettles and I identified the problem event of 2003 near China's Lop Nor test site as an earthquake by five different techniques.

Nettles and I also identified (not shown) several large mine collapses in various countries of magnitude 2.8 to 5. They are typically richer in long-period seismic waves than earthquakes, making their identification as distinct from underground nuclear explosions even easier. Mechanism solutions for some of those events also indicate that they were collapses and not explosions. Several very large chemical explosions were detected and identified as well.

In summary, the detection and identification of problem seismic events of several kinds—earthquakes, nuclear explosions, alleged nuclear explosions, mine collapses, and chemical explosions—have improved dramatically over the past five decades. A residuum of less than one problem event per year in countries of concern to the United States can be reduced further with special studies.

15

SENATE REJECTION OF THE CTBT IN 1999

President Clinton signed the Comprehensive Nuclear Test Ban Treaty (CTBT) along with a number of other world leaders at the United Nations in September 1996. When it came up for a vote in 1999, the U.S. Senate failed to ratify the treaty. Nevertheless, it still remains on the Senate's agenda. Ratifying a treaty requires the Senate to pass it by a two-thirds vote and the president to then sign it. Those countries that have signed but not ratified it are nonetheless still bound by the Vienna law of treaties to obey the terms of the CTBT. None of the signers has violated the terms of the CTBT by conducting nuclear tests since September 1996. This was a considerable accomplishment.

ACTIONS PRIOR TO SENATE DEBATE

Several developments in 1997 and 1998 preceded the Senate's vote on the treaty. On September 22, 1997, President Clinton submitted the CTBT to the Senate for what is termed its advice and consent. He also included the following six safeguards with his submission:

1. A Science-Based Stockpile Stewardship program to ensure a high level of confidence in the safety and reliability of U.S. nuclear weapons in the active stockpile, including the conduct of a broad range of effective and continuing experimental programs.
2. The maintenance of modern nuclear laboratory facilities and programs in theoretical and exploratory nuclear technology that will attract,

retain, and ensure the continued application of our human scientific resources to those programs on which continued progress in nuclear technology depends.

3. The maintenance of the basic capability to resume nuclear test activities prohibited by the CTBT should the United States cease to be bound to adhere to this treaty.
4. The continuation of a comprehensive research and development program to improve our treaty monitoring capabilities and operations.
5. The continuing development of a broad range of intelligence gathering and analytical capabilities and operations to ensure accurate and comprehensive information on worldwide nuclear arsenals, nuclear weapons development programs, and related nuclear programs.
6. The understanding that if the President of the United States is informed by the Secretary of Defense and the Secretary of Energy (DOE)—advised by the Nuclear Weapons Council, the Directors of DOE's nuclear weapons laboratories, and the Commander of the U.S. Strategic Command—that a high level of confidence in the safety or reliability of a nuclear weapon type that the two Secretaries consider to be critical to our nuclear deterrent could no longer be certified, the President, in consultation with the Congress, would be prepared to withdraw from the CTBT under the standard "supreme national interests" clause in order to conduct whatever testing might be required.

Similar safeguards had been adopted when the Limited Test Ban Treaty was submitted to the Senate in 1963.

In September 1997, Senator Joseph Biden made a major speech emphasizing the value of the treaty to U.S. national security. He said, "The time has come, Mr. President, to move ahead on the Comprehensive Test Ban Treaty, as well as other arms control initiatives and NATO enlargement." Several other senators spoke in favor of the CTBT later that fall.

The *Congress Daily* reported on November 25, 1997, "Key Senator, Pete Domenici, Hesitates In Supporting CTBT" . . . A key Republican senator this week said his support for a nuclear test ban treaty depends on increased funding for federal energy laboratories, two of which are located in his home state of New Mexico." Richard Garwin, a physicist who has long been involved in the control of nuclear weapons, and

I visited Domenici's Senate office shortly afterward to discuss the seismic event of August 1997 and to assure him that it was not a Russian nuclear explosion. Nevertheless, Domenici followed party discipline and voted against the treaty in 1999.

In February 1998, President Clinton visited the Los Alamos nuclear weapons lab, where he told employees, "The test ban treaty will hold other nations to the same standards we already observe." He visited a day after proposing additional funding to the weapons labs for work to assure that the U.S. stockpile of nuclear weapons would work in the future without nuclear testing. That program, Stockpile Stewardship, called for $4.5 billion per year to be spent at the labs on the fastest computers in the world, new major high-energy facilities to retain capabilities in nuclear science, and better documentation of existing weapons and past tests. Clinton also proposed a package of safeguards to expand intelligence and the monitoring of other countries' nuclear programs. He pledged that the United States would reserve the right to resume nuclear tests if the safety of the nuclear stockpile could no longer be certified.

Clinton's trip to New Mexico, an attempt to pacify hard-line opponents, was regarded as a respite from Washington, where he faced a barrage of questions about his sexual affair with Monica Lewinsky. That affair and the subsequent attempt to impeach him undoubtedly negatively affected many of his proposals, including ratification of the CTBT. The Senate's negative vote in 1999 has aspects of attacking Clinton when he was weak politically. Some people commented that all Republicans hated Clinton whereas only half hated the Comprehensive Nuclear Test Ban Treaty.

In May 1998, Clinton criticized India and Pakistan soon after their nuclear tests, citing the dangers they posed to one another and to the rest of South Asia. In July 1999, he congratulated Brazil on its ratification of the CTBT. Clinton stated, "Brazil's action today to ratify the CTBT makes it all the more important for the United States to do the same. I call on our Senate to act expeditiously to approve the CTBT—already signed by 149 nations and supported by the Joint Chiefs of Staff—so that the United States can lead in this vital endeavor."

Republicans controlled the Senate in 1998 as well as during the debate and vote on the CTBT in 1999. In the wake of India's and Pakistan's nuclear tests of May 1998, Senate majority leader Trent Lott and Foreign Relations Committee chair Jesse Helms claimed that a ban on nuclear testing was "irrelevant" and that the Senate should not even debate the CTBT. Helms served notice that he would not move toward ratification of the CTBT until amendments to the Anti–Ballistic Missile (ABM) Treaty were debated. He hoped the ABM Treaty would be defeated by the Senate. Lott also suggested that U.S. policy should shift from pressing for a CTBT to the construction of "effective missile defenses," claiming that "the Administration's push for the CTBT led to the [Indian and Pakistani] nuclear tests." Helms also called for the defeat of the Kyoto agreement on global warming.

Many articles supporting the CTBT were published in the United States during 1999. Several senators, scientists, and arms control specialists refuted the views of Lott and Helms in the media and in the Congress. Among them was an op-ed article in the *Washington Post* on June 11 titled "This Treaty Must Be Ratified," by Sidney Drell of Stanford and Paul Nitze. Nitze was a former arms control negotiator and an ambassador-at-large in the Reagan administration. They noted that other arms control treaties had been brought before the Senate, whereas that body had failed to consider the CTBT for two and a half years after Clinton signed it at the UN. They commented, "The president rightly has referred to the CTBT as the 'longest-sought, hardest-fought prize in the history of arms control.'"

Former Secretary of State Henry Kissinger stated at a hearing chaired by Senator Helms on May 26, 1999, that he was positive on missile defense. Senator Frist then asked him about the test ban, noting the "constraints" it would place on the United States. Kissinger said, "I think we have an arms control objective, and must have, to prevent the proliferation of nuclear weapons. And anything that makes it more difficult to develop more nuclear weapons, I would, in principle, favor." He later said he thought the CTBT was a poor treaty.

Daryl Kimball of the Physicians for Social Responsibility became the executive director of a task force on the CTBT, called the Coalition to Reduce Nuclear Dangers. It represented a number of arms control organizations in the United States. On June 15, 1999, Kimball sent a letter to President Clinton signed by thirty-five heads of organizations and former government officials urging a substantial presidential campaign on behalf of the CTBT. The letter asked the president to do the following:

1. Present your case for CTBT ratification directly to the public and invite bipartisan support for its consideration and approval on a frequent and consistent basis.
2. Appoint a high-level, full-time CTBT coordinator to strengthen and focus administration-wide efforts and to signal the seriousness with which you plan to pursue ratification.
3. Direct key cabinet members and high-profile CTBT supporters to pursue a sustained, public campaign to increase support and win Senate approval for the treaty.

While Clinton himself seemed to be familiar with the main test ban issues during the period leading up to the Senate vote, his administration unfortunately did not adopt any of these measures. Likewise, the Obama administration did not adopt any of the three suggestions despite making general statements about the need for the Senate to ratify the treaty and its value to national security.

Many proponents of the CTBT cited polls indicating wide public support in the United States for the treaty, some as high as 82 percent. In a speech on July 27, 1999, Senator Biden, the ranking minority member on the Senate Foreign Relations Committee, said, "Listen to the American People. I know that some of my colleagues have principled objections to this treaty. I respect their convictions, even though I strongly believe they are wrong on this issue. Some of my colleagues believe nuclear weapons tests are essential to preserve our nuclear deterrent. Both I and the directors of our three nuclear weapons laboratories disagree. The $45 billion [for ten years] 'Stockpile Stewardship' program enables us to maintain the safety and reliability of our nuclear weapons without weapons tests." He also noted, "Support for ratification is not limited, moreover, to the current Chairman of the Joint Chiefs of Staff. The four previous Chairmen also support ratification."

The Office of Technology Assessment (OTA) was established in 1972 to provide Congress with assessments of various scientific and technical proposals that were relevant to hearings and legislation. OTA was defunded during the period of Newt Gingrich's Republican majority rule in the House of Representatives, often referred to as the "Contract with America." OTA closed in September 1995. Its demise was a blow to further scientific and technical assessments of the CTBT and other arms control issues. In July 2010 the House voted 235 to 176 to reject a proposed amendment to restore OTA.

The *New York Times* reported on August 30, 1999, that Democrats were threatening to bring the Senate to a standstill when Congress returned the following month from summer recess unless Republicans agreed to hold hearings that year on the Comprehensive Test Ban Treaty. The article quoted Senator Biden: "But the President has to play a major role. He could affect this more than he has." The *Times* went on to state, "Administration officials say that in the coming weeks, Clinton and top foreign policy aides, like Secretary of State Madeleine K. Albright and national security adviser, Samuel R. Berger, will push the treaty more publicly."

SHORT SENATE HEARINGS AND DEBATE

Many Democrats and other supporters assumed the treaty would readily pass the Senate if Helms and Lott released their lock on it. Unfortunately, that was not the case. As Senate majority leader, Lott had major control over bringing bills and treaties to the floor of the Senate. On September 30, he suddenly proposed short hearings, ten hours of debate, and a vote as early as October 5 on the CTBT. Early October gave the Clinton administration too little time to push more actively for the treaty. Lott claimed that he had the votes to defeat the treaty, which soon turned out to be correct.

Senators Helms and Lott argued on the Senate floor on October 1, 1999, that the Foreign Relations Committee had held hearings in which the treaty was mentioned. They offered a *unanimous consent resolution* [my italics] for a vote on final passage of the CTBT on October 7, 1999. Unanimous consent requires that all senators present vote for such a resolution. In an exchange in the Senate, Democratic minority leader Thomas Daschle objected to Lott's offer, saying it gave short shrift to a landmark pact. He requested more time for debate and a later date for the vote.

Some senators, however, urged Daschle to accept the offer to bring it to a vote. Senator Biden told reporters, "This is not the best way to deal with this, but it's better than nothing." He urged the president to "pull out all the stops" and to hold a news conference and a nationally televised address. Some senators feared that a vote now would doom the treaty, while others said they were ready to "roll the dice," even if they lacked the votes. A number thought that votes against the treaty would reflect badly on Republican senators who were up for reelection in November, but this proved not to be the case. In retrospect, the vote to defeat the treaty a week later turned out to be worse than nothing.

On October 1, Senators Lott and Daschle agreed to a revised unanimous consent resolution that included fourteen hours of general Senate debate on the CTBT, starting on October 8 and resuming on October 12, with a vote taking place that day or soon thereafter. Every Democrat in the Senate endorsed the timing, a major mistake. The Democrats, probably unaware of how dire their situation was, were trapped by their own rhetoric. How could they back out now? Senator Jon Kyl, a treaty opponent said, "It was plain arrogance. . . . They didn't have any idea they wouldn't win."

Attempts were made soon after passage of that resolution to postpone the vote and to have more time for testimony and debate. Senators John Warner, a Virginia Republican, and Patrick Moynihan, a New York Democrat, released a letter to their two leaders, endorsing efforts to delay Senate action on the treaty until the next Congress convened in 2001. About a dozen senators signed the letter. Most Democratic senators and the president, though, seemed to have been caught off guard about changing the unanimous consent resolution. It was soon clear that senators James Inhofe and Jon Kyl, who wanted to defeat the treaty, would not accept postponements. Hence, unanimous consent could not be reached to postpone the vote, and it went ahead.

The *New York Times* reported the day after the vote, "In a last-ditch effort to save the treaty, Clinton called the Republican leader, Trent Lott, two hours before the vote and asked that he delay action for national security reasons. In a blunt rebuff, Lott said the President had offered too little, too late, and he pushed ahead with an action that he knew would humiliate Clinton."

In the weeks before the vote, I talked several times by phone with Daryl Kimball, who headed the CTBT task force of nongovernmental organizations, and suggested that he enlist the support of Republican moderates, such as the two senators from Maine. He thought that was not needed. In retrospect, it is clear that many senators who were thought to be undecided had already made up their minds to vote against the CTBT. Democrats and other supporters of the treaty were unaware of that development.

During the brief Senate debate, lack of verifiability and concern about the U.S. stockpile of nuclear weapons were cited repeatedly. Not emphasizing verification surprised many of us because an elaborate International Monitoring System (IMS) had been set up under the 1996 treaty. In addition, the United States possessed many other monitoring tools, so-called National Technical Means (NTM).

The negative views should have come as no surprise given repeated claims by treaty opponents stretching back to 1954 that other nations, particularly the Soviet Union and China, would cheat and the United States would not be able to detect their nuclear tests, especially those conducted evasively. I said in my *FAS* article, "Concerns about verifiability as well as the reliability of weapons in the U.S. stockpile, in fact, have long served as proxies for the larger issues of what best ensures U.S. national security and prevents nuclear war."

On October 3, 1999, the *Washington Post* published on its front page one of the most damaging articles to ratification of the treaty by Robert Suro. In "CIA Is Unable to Precisely Track Testing," he stated that the CIA had concluded that it could not monitor low-level nuclear tests by Russia precisely enough to ensure compliance with the CTBT. He went on to say that twice in September 1999, "the Russians carried out what might have been nuclear explosions at its Novaya Zemlya testing site in the Arctic." Bill Gertz of the *Washington Times* also published allegations of nuclear tests at Novaya Zemlya in 1999 just as the CTBT debate was heating up. Gertz's comments at a critical time in the CTBT debate have much the same flavor as his claims that the Kara Sea earthquake of August 1997 was a Russian nuclear explosion, which it clearly was not. Suro said congressional staffers [all of whom worked for Republicans] were briefed on the new CIA assessment before Lott scheduled the vote on the CTBT.

Those giving false information to Suro obviously had little knowledge of how well Novaya Zemlya was being monitored.

In my *FAS* article of March 2000 I stated, "We know now that Senators Cloverdell and Kyl, who strongly opposed the Treaty, and their staffs worked in secret for months to compile briefing books of materials opposing the CTBT before Lott's sudden announcement. Marshall Billingslea of Senator Helm's staff stated he worked exclusively for two years on arguments to defeat the Treaty. His materials, which were not reviewed, were made available to Senators likely to vote against the CTBT but not to Democrats or other Republicans who were either likely to vote in favor or leaned toward ratification. Letters opposing the Treaty that were obtained from several former national security and defense officials were cited repeatedly in the Senate debate." Helms and Lott worked long and hard in secret to gather votes against the CTBT. In essence, they conducted a successful ambush.

Although officials in the Clinton administration cited the CTBT as one of the president's top foreign policy priorities, little was done until after Lott's announcement to aggressively promote the treaty in the Senate, especially among moderate Republicans, or to describe its main benefits. No high-level official in the executive branch was designated to promote and organize support for the CTBT. Most proponents of the treaty only began active efforts on its behalf a few days before the short Senate hearings.

The hearings before Senate committees did not do justice to accomplishments in stockpile stewardship and especially to nuclear verification. Physicists Sidney Drell and Richard Garwin spoke about the stockpile. Only one witness, General John Gordon, deputy director of the CIA, testified about verification, and that was in a closed secret session. I am not aware that he was an expert on verification, especially seismic identification and evasive testing. My colleague Paul Richards wrote to me, "Current unclassified methods monitor that site [Novaya Zemlya] down to about 0.01 kt. It would not surprise me to learn that a much larger number was indicated to Sen. Warner's committee last Tuesday."

Drell stated, "This treaty can be effectively verified. With the full power of its international monitoring system and protocols for on-site inspection, we will be able to monitor nuclear explosive testing that might undercut our own security in time to take prompt and effective counteraction." In contrast, Paul Robinson, director of the Sandia National Laboratory, testified, "If the United States scrupulously restricts itself to zero-yield while other nations may conduct experiments up to the threshold of international detectability, we will be at an intolerable disadvantage." Intolerable is a strong word. Robinson, in fact, is not an expert on verification but on the design of weapons. It is highly likely he was not aware of how good verification was in 1999.

Robinson continues his long opposition to a CTBT today, even though detection and identification thresholds have continued to get much better. When seismic and other monitoring technologies improved significantly at various times during the past forty years, Robinson and others claimed in hearings that we still must do better. While Robinson and some others apparently believed that a few kilograms of nuclear explosive yield would be of great military advantage to some countries, other scientists with similar high-level clearances strongly disagreed.

On October 8, 1999, the three directors of the U.S. nuclear weapons labs issued the following joint statement about the U.S. stockpile: "We, the three nuclear weapons laboratory directors, have been consistent in our view that the stockpile remains safe and reliable today. For the last three years, we have advised the Secretaries of Energy and Defense through the formal annual certification process that the stockpile remains safe and reliable and that there is no need to return to nuclear testing at this time. We have just forwarded our fourth set of certification letters to Energy and Defense Secretaries confirming our judgment that once again the stockpile is safe and reliable without nuclear testing. While there can never be a guarantee that the stockpile will remain safe and reliable indefinitely without nuclear testing, we have stated that we are confident that a fully supported and sustained stockpile stewardship program will enable us to continue to maintain America's nuclear deterrent without nuclear testing. If that turns out not to be the case, Safeguard F—which is a condition for entry into the Test Ban Treaty by the U.S.—provides for the President, in consultation with the Congress, to withdraw from the Treaty under the standard 'supreme national interest' clause in order to conduct whatever testing might be required."

In 1999 Fred Eimer of the U.S. Arms Control and Disarmament Agency stated that, by presidential directive, an effective verification system "should be capable of identifying and attributing with high confidence evasively conducted nuclear explosions of about a few kilotons yield in broad areas of the globe." That yield is far larger than that of a hydronuclear explosion with a nuclear yield of a few kilograms—that is, a few millionths of a kiloton. In his Senate testimony in 1999, Garwin stated, "Without nuclear tests of substantial yield, it is difficult to build compact and light fission weapons and essentially impossible to have any confidence in a large-yield two-stage thermonuclear weapon or hydrogen bomb. Can one be certain that a nation has not tested in the vast range between zero and the magnitude of tests that would be required to gain significant confidence in an approach to thermonuclear weaponry—say, 10 kilotons? No, but the utility of such tests to a weapons program has been thoroughly explored and found to be minimal."

On October 8, 1999, during the debate, Senator Lott cited many reasons for opposing the CTBT. On verification he said, "We know, however, that it is possible to conduct a nuclear test with the intention of evading systems designed to detect the explosion's telltale seismic signature. This can be done through a technique known as 'decoupling,' whereby a nuclear test is conducted in a large underground cavity, thus muffling the test's seismic evidences." Lott went on to state, "In a speech to the Council on Foreign Relations last year, Dr. Larry Turnbull, Chief Scientist of the Intelligence Community's Arms Control Staff, said, 'The decoupling scenario is credible for many countries for at least two reasons: First, the worldwide mining and petroleum literature indicates that construction of large cavities in both hard rock and salt is feasible, with costs that would be relatively small compared to those required for the production of materials for a nuclear device; second, literature and symposia indicate that containment of particulate and gaseous debris is feasible in both salt and hard rock.'"

Lott added, "So not only is this 'decoupling' judged to be 'credible' by the Intelligence Community, but according to Dr. Turnbull, the technique can reduce a nuclear test's seismic signature by up to a factor of 70. This means that a 70-kiloton test can be made to look like a 1-kiloton test, which the CTBT monitoring system will not be able to detect."

Senator Helms made similar remarks about evasion with a 60-kiloton nuclear test. He said, "Every country of concern to the U.S.—every one of them—is capable of decoupling its nuclear explosions . . . without detection by our country."

Lott also cited James Woolsey, Clinton's first CIA director, "I do not believe that the zero level is verifiable. Not only because it is so low, but partially because of the capability a country has that is willing to cheat on such a treaty, of decoupling its nuclear tests by setting them off in caverns or caves and the like." No one challenged these claims during the very short Senate hearings and debate. The general sense among the senators and their staffs who opposed the treaty was "just put it in a hole" and you can defeat the monitoring system. Examples of large underground facilities, such as the Washington, DC, metro stations, created the impression that such an undertaking was well within any group's capability.

These claims were in stark contrast to a joint public statement by the American Geophysical Union and the Seismological Society of America on October 6, 1999. It said in part, "The decoupling scenario, however, as well as other evasion scenarios, demand extraordinary technical expertise and the likelihood of detection is high. AGU and SSA believe that such technical scenarios are credible only for nations with extensive practical testing experience and only for yields of at most a few kilotons. Furthermore, no nation could rely upon successfully concealing a program of nuclear testing, even at low yields."

Woolsey had long been an opponent of a CTBT and other arms control agreements. His claim about setting off decoupled nuclear tests successfully in caves was incorrect. As already discussed in chapter 4, caves are typically too shallow to contain the radioactive particles and gases produced by even very tiny nuclear explosions. Large and deep caves are typically filled with water and would be difficult, if not impossible, to pump dry. Explosions in a water-filled cave or cavity are very well coupled, not muffled. Most caves have many openings to the atmosphere, some of which may not be known to a potential evader. A proposal to use an old salt mine at Lyons, Kansas, for the long-term storage of high-level radioactive waste was abandoned because all of the openings to the atmosphere could not be identified with confidence.

Turnbull's assertions cited by Senator Lott about decoupled testing—its feasibility, containment, and detection—are false for yields larger than one kiloton. Claims by Lott and Helms in 1999 that decoupled explosions of 60 and 70 kilotons are even feasible, let alone undetectable, were huge exaggerations.

Turnbull's opinions and great negative influence on the test ban debate were well known. He revealed his title at the CIA as early as 1992 in the published program for an international conference at Princeton on nuclear verification. I have known him, his professional background, and his views for decades and strongly believe that he should have been fired decades ago for his incompetence. One of my professional colleagues observed, "The bottom line is that in an important sense Turnbull is right. The U.S. cannot do a good job of CTBT verification. The reason is that people like Turnbull are in charge."

Among those who supported the treaty were French president Chirac, British prime minister Blair, and German chancellor Schroeder. They coauthored an op-ed article in the *New York Times* on October 8, 1999, in which they stated, "Failure to ratify the treaty will be a failure in the struggle against nuclear proliferation." They contended the test ban treaty was verifiable.

Kidder, von Hippel, and I wrote an op-ed article, "False Fears About a Test Ban," in the *Washington Post* on October 10, 1999, during the Senate debate. We said in part, "More than 80 percent of the American people want a permanent ban on nuclear weapons tests, and support outside the United States is at least as high. This public support has powered the movement that persuaded the governments of 154 nations to sign a Comprehensive Test Ban Treaty. The arguments against the test ban treaty today are the same as those that opponents used to slow its progress for 40 years: the fear that other countries will cheat and be able to reap advantages from small clandestine tests, and the belief that the only way to make sure that a nuclear weapon works is to test it.

"Still there is a possibility that a small nuclear test, carried out secretly away from monitored test sites, might escape detection. But what could be gained from such a test? Very little could be gained below the threshold for the 'boosting' of fission explosives. [Boosting] the yield with the fusion of a small amount of tritium-deuterium gas was the key step in the development of modern compact warheads, a 'secret' that has been officially declassified for decades. Testing boosting requires a nuclear explosion with a power of at least a few hundred tons of TNT, and full boosting gives yields of thousands of tons. This is beyond the level that could plausibly be concealed from U.S. seismic monitoring stations. The United States has done almost no testing for nuclear weapons development below 1,000 tons of TNT, so we can be comfortable with a ban on nuclear tests of all sizes.

"The analyzed classified record shows that since the U.S. nuclear establishment mastered the art of designing boosted thermonuclear weapons more than two decades ago, there have been virtually no failures. On the surface, the debate over the nuclear test ban is about technical uncertainties. Below the surface, it is about competing priorities. Many test ban opponents care only that the United States be unconstrained in the development of nuclear weapons. If this country resumed testing, however, other countries would as well. They would improve their nuclear weapons much more than we would and the world would be pushed back closer to nuclear weapons use."

Senator Kyl of Arizona spoke first against the Comprehensive Test Ban Treaty during the Senate debate, which started on October 8. He said that if the treaty were passed, the United States would not be able to test to assure the safety and effectiveness of American nuclear weapons. He also said, "Our intelligence agencies lack the ability to confidently detect low-yield test[s]."

On the same day, President Clinton, visiting Canada, said, "I hope that the Senate will reach an agreement to delay the vote and to establish an orderly process, a non-political orderly process, to systematically deal with all the issues. . . . we've been trying to have a debate on this for two years, but it is clear now that the level of opposition to the treaty and the time it would take to craft the necessary safeguards to get the necessary votes are simply not there." I think he should have been in Washington during the CTBT debate. Nevertheless, three officials in his administration spoke at length at a news conference about the advantages of a CTBT. They did not cover verification or the possibilities of cheating very well.

Former secretary of defense and former CIA director James Schlesinger called those who wanted to move away from nuclear weapons unrealistic "abolitionists" who did not grasp the importance of maintaining a reliable deterrent for decades to come. Schlesinger, who testified against the treaty and has long been an opponent of test bans, said it might not pose a substantial threat to American security for some years, "but the real questions come after about 2020," when America's overwhelming conventional military superiority may have eroded and new players on the world stage could threaten the United States. If the U.S. military were locked into a permanent test ban treaty, Schlesinger complained, its options could be dangerously limited. He argued that it was vital to retain the right to test and improve the U.S. nuclear arsenal. Only actual test explosions, he said, could confirm the dependability of the American deterrent and convince both friends and foes in decades to come that U.S. nuclear weapons remained operational. The high-level positions he held made him a strong voice in defeating the CTBT. He died in March 2014.

Writing in the *Washington Post* on October 13 just before the Senate vote, Robert Kaiser said that Brent Scowcroft, national security adviser to President George H. W. Bush and a weathered veteran of the arms control wars, complained, "We still think of arms control in Cold War terms. We have not changed at all, and yet the world has changed dramatically." Kaiser continued, "The debate in the Senate, Scowcroft added, 'is pathetic.' Scowcroft said he hoped for a debate some day that would connect the test ban treaty to a realistic assessment of what the United States can do to ensure 'security and stability.' That means looking carefully at how various policies will affect the thinking of potential rivals or potential nuclear powers, and deciding what kind of U.S. arsenal best suits national interests. Scowcroft had formally recommended that the Senate put the treaty aside. 'Let's wait until after this partisan period has passed and we can debate it sensibly,' he said, hopefully."

The treaty went down to defeat on October 13, 1999, by a majority vote of fifty-one to forty-eight, far fewer than the two-thirds, or sixty-seven votes, needed for the Senate's advice and consent. It was largely, but not completely, a party-line vote, with forty-four Democrats voting for the treaty. No Democrats voted no, but Senator Byrd voted present. Republicans Chafee, Spector, Jeffords, and Smith voted for the CTBT. Jeffords and Spector, moderate Republicans, each gave long, thoughtful speeches and wrote op-ed articles on the value of the treaty. Jeffords later switched to the Democratic Party.

POSTMORTEM

The CTBT is likely the most important treaty to be defeated since the Senate's vote against the Versailles Treaty in March 1920 and against U.S. participation in the League of Nations. The day after the Senate's defeat of the CTBT, presidential candidate Al Gore, then the vice president, vowed to push for its ratification.

Secretary of State Madeleine Albright announced on November 11, 1999, that the Clinton administration "will establish a high-level Administration task force to work closely with the Senate on addressing the issues raised during the test ban debate." While Albright reiterated many of the arguments for a CTBT, she and others said nothing about the leaks of information to the press about alleged nuclear tests by Russia and China prior to and during the Senate debate. A strong rebuttal was needed, but not given, to the *Washington Post* article of September 1999 stating that the CIA had concluded it could not monitor low-level nuclear tests by Russia precisely enough to ensure compliance with the CTBT.

I have saved electronic copies of many letters to editors, communications among treaty supporters, speeches on the Senate floor, and copies of a letter I sent in November 1999 to Senators Biden, Levin, and Moynihan about misleading and incorrect statements about verification during the CTBT debate of 1999. They will be deposited at the Rare Book and Manuscript Library of Columbia University. Speeches in the Senate can be found in the *Congressional Record, Senate.*

16

THE CTBT TASK FORCE AND THE 2002 AND 2012 REPORTS OF THE NATIONAL ACADEMIES

THE CTBT TASK FORCE

In January 2000, U.S. Secretary of State Madeleine Albright announced that former chairman of the Joint Chiefs of Staff, retired General John Shalikashvili, had agreed to serve as an adviser to President Clinton and to her in order to spearhead the administration's effort to achieve bipartisan support for ratification of the Comprehensive Nuclear Test Ban Treaty (CTBT) after its defeat in the U.S. Senate the year before.

Shalikashvili agreed to reach out to senators to find ways to narrow differences over the CTBT. He met for the first time in March 2000 with his CTBT task force. In a speech that month he stated, "I remain convinced that the United States will be safer with this important treaty than without it. True, potential proliferators can make simple fission bombs without testing. But a test ban makes it much harder to get nuclear weapons down to the sizes, the shapes and the weights most dangerous to us: deliverable in light airplanes, rudimentary missiles, or even in a terrorist's luggage."

On April 20, Shalikashvili indicated that it might well be necessary to attach conditions and understandings to the U.S. signing statement for the CTBT to convince senators to agree to ratification of the pact. The main concerns that he found after meeting with various senators who had voted against the CTBT in 1999 were: (1) the Treaty should not be of indefinite duration; (2) the United States should wait until the Stockpile Stewardship Program is completed; and (3) a treaty with a zero yield cannot be verified. He commented that he did not expect the Senate to act on the treaty in the remainder of 2000 and that his goal was to work toward developing "a more reasoned judgment" in the next administration.

FIRST STUDY BY THE NATIONAL ACADEMIES

To support his efforts, General Shalikashvili commissioned several studies, including one started in April 2000 with the U.S. National Academies to address major technical issues that had arisen during the 1999 Senate debate. Its mandate was confined to a specific set of important technical questions. The Academies were not asked to provide an overall "net assessment" of whether the CTBT was in the national security interest of the United States, which they did not do. The official government sponsor of the first study was the Department of State, with additional funding from the Department of Energy, the National Academy of Sciences (NAS), and several U.S. foundations.

For this study and another a decade later, the NAS turned to its standing Committee on International Security and Arms Control (CISAC), created in 1980. In 2000 CISAC picked members of the first CTBT study, with John Holdren of Harvard University as chair. Holdren, who later became President Barack Obama's science adviser, had been involved for many years in arms control and energy issues. The committee contained members with special expertise, including those from academia, high-ranking retired military officers, persons involved in past arms control negotiations, and former officials of the weapons laboratories. The members also were chosen with regard for appropriate balance, including geophysicist Raymond Jeanloz from UC Berkeley and my seismological colleague Paul Richards from Columbia University.

After the committee signed off on a draft document in December 2000, its report, *Technical Issues Related to the Comprehensive Nuclear Test Ban Treaty*, then entered an extended process of multiagency classification review and peer review by the National Academies. The committee held extensive unclassified and classified meetings. General Shalikashvili was briefed by the committee at the classified level prior to the end of Clinton's presidency. An unclassified report with a classified annex was published in 2002 during the Bush administration.

The report addressed three main concerns:

1. The capacity of the United States to maintain confidence in the safety and reliability of its nuclear stockpile in the absence of nuclear testing
2. The capabilities of the international nuclear test monitoring system and possibilities for decoupled nuclear explosions
3. Additions to their nuclear weapons capabilities that other countries could achieve through nuclear testing at yield levels that might escape detection—as well as the additions they could achieve without nuclear testing at all—and the potential effect of such additions on the security of the United States

By 2002 considerable advances had been made in stockpile confidence, nuclear monitoring, and dealing with evasive testing. The headline of the press release for the report published in 2002 was "Verification Capabilities Are Good, Cheating Possibilities Are Limited, and Safety and Reliability of U.S. Weapons Can Be Maintained Without Nuclear Tests."

The 2002 report stated that a weapon's reliability is dominated by the nonnuclear components of the entire system, such as electronics, which are testable under a CTBT. Stockpile stewardship by means other than nuclear testing is not a new requirement imposed by the CTBT; it has always been the mainstay of the U.S. approach to maintaining confidence in stockpile safety and reliability. Since 1996, much more had been learned under the Stockpile Stewardship Program about the stability of plutonium on time scales of many decades. Its stability is crucial to the functioning of the primary (fission) stage of thermonuclear weapons.

New data for nuclear monitoring had become available by 2002 when the International Monitoring System (IMS) was only partly in place. In addition, modern digital seismic data became available in near real time from hundreds of other global stations that were not part of the IMS. The capacity of computers to handle huge amounts of data, especially seismic waveforms, increased tremendously as well. Monitoring Russia and China became possible by 2002 once data started flowing from stations within those countries, something the United States had long considered essential for monitoring a full test ban. Data also became available from countries surrounding Russia and China, such as Mongolia, Kazakhstan, and Kirgizstan.

The new government of Kazakhstan worked to destroy tunnels and shafts at the former Soviet nuclear test site in the eastern part of its country. As part of that program, Kazakhstan detonated three large chemical explosions at varying depths to explore how much their seismic magnitudes changed with depth. All three were detected at stations of the International Monitoring System as far away as 4600 miles (7500 km).

The 2002 report concluded, "Taking all factors into account and assuming a fully functional IMS, we judge that an underground nuclear explosion cannot be confidently hidden if its yield is larger than 1 or 2 kilotons." This was the capability I had claimed in my 1996 paper "Dealing with Decoupled Nuclear Explosions Under a Comprehensive Nuclear Test Ban Treaty."

William Leith of the U.S. Geological Survey and I debated the feasibility of conducting hidden decoupled nuclear tests of various yields at one meeting of the committee on August 10, 2000. Leith's expertise was based almost solely on his unpublished global catalog of very large holes in the ground. He argued before the CTBT committee and in a USGS Open File Report in 2001 that huge underground holes could be used for evasive decoupled testing of nuclear devices of 10 kilotons and larger.

Leith's catalog, however, listed many huge holes that were open to the atmosphere, such as a giant sink (karst) hole in the jungles of Indonesia, which were not suitable for clandestine nuclear testing. Other buried openings on his list, such as the Norwegian skating rink for the Olympic games, were so shallow that their tops would be blown off even by very small nuclear explosions. Nevertheless, his views were repeated and quoted by Cyrus Knowles, Don Linger, and others of the former Defense Nuclear Agency and Larry Turnbull of the Arms Control Intelligence Staff. Fortunately, the relevant part of the 2002 NAS report, which I did not write, accepted my views on decoupling and not those of Leith.

A summary conclusion of the 2002 report stated that to get the large efficiency gains and weight reductions associated with boosting, an inexperienced state would need to test repeatedly at yields well above a kiloton, which it would not be able to conceal reliably. It also noted that considerable weapon design experience would be required to achieve low yields.

OTHER DEVELOPMENTS IN 2000

By mid-2000, all members of NATO except the United States had ratified the CTBT. In June 2000, the Russian Federation also ratified it, along with the bilateral strategic arms limitation treaty, START II, after seven years of delays. That treaty, ratified by the U.S. Senate in 1996, committed each side to cut its nuclear arsenal down to between 3000 and 3500 warheads, or about half the number allowed by START I. When he became president, George W. Bush gave notice to Russia that the United States would no longer adhere to the Anti–Ballistic Missile (ABM) Treaty and would build an ABM system; Russia then withdrew from START II.

On July 19, 2000, I attended the Stanford Center for International Security and Cooperation–Lawyers Alliance for World Security roundtable discussion on the CTBT in Palo Alto, California. General Shalikashvili attended, along with Republican senator Charles Hagel of Nebraska. Although Hagel had voted against the CTBT in 1999, he thought that much more time and debate should have been devoted to the treaty. In 2013 and 2014, he was secretary of defense. He and Shalikashvili showed great interest in what was said at the meeting about verification and stockpile stewardship.

Meanwhile, in early 2000, Sandia Lab Director Paul Robinson continued to claim difficulties in maintaining nuclear weapons without tests and to advocate the development of new nuclear weapons. He said that not doing so was tantamount to a policy of "self-deterrence" by the United States, in which the country would be giving up flexibility to respond to crises in a world with many nuclear powers. The Sandia National Laboratories is now a wholly owned subsidiary of the Lockheed Martin Corporation. Its primary mission is to develop, engineer, and test the nonnuclear components of nuclear weapons.

On March 29, 2000, the *Albuquerque Journal* reported that Robinson's assertions were challenged by retired vice president of Sandia, Bob Peurifoy. Peurifoy said the weapons labs simply needed to focus on their mission of maintaining existing weapons—weapons they knew worked and would work for a long time. He suggested that what each of the weapons labs needed most was "a chief engineer" whose only mission would be assessing "the health of the stockpile." Peurifoy said he had nothing against putting expensive bomb simulators and supercomputers at each of the nuclear weapons labs, if taxpayers were willing to pay for them, but commented that lab directors were misleading Congress and the public about the need for them.

A continuing fear is that funding of current programs to inspect existing weapons and to manufacture and replace aging parts will suffer as greater amounts of funding are devoted to the National Ignition Facility at Livermore and other very expensive long-term facilities at the weapons labs.

On May 31, 2000, Terry Wallace of the University of Arizona, Gregory van der Vink, and I convened a special session "The Comprehensive Test Ban Treaty: Issues of Verification and Monitoring" at the meeting of the American Geophysical Union in Washington, DC. Acting Assistant Secretary of State O. J. Sheaks opened the session with a broad overview of the CTBT, focusing on its role in promoting global arms control and nonproliferation objectives. Several of us described how poorly the Senate had examined the CTBT, especially verification and monitoring. I described the negative role of Larry Turnbull as cited by senators Lott and Helms in 1999.

On June 12, 2000, the *Washington Post* reported that new nuclear weapons were on the minds of a small but powerful cadre in the United States. It said that Senate Republicans had put a provision in the FY 2001 defense authorization bill that specifically required the secretaries of defense and energy to undertake a study to develop a new "low-yield" nuclear weapon that could destroy deeply buried targets and to permit the nuclear weapons labs to conduct limited research and development that might be necessary to complete the study. If it had become law, which it did not, it could have provided a rationale for resuming nuclear testing to confirm the new warhead design.

By April 2000 it was clear that Bush and Gore would be their parties' presidential candidates in the November election. Gore was in favor of the CTBT. Bush was against it and remained so during his two terms in office. Bush, however, advocated and continued a moratorium on nuclear testing. One member of his administration publically advocated a resumption of nuclear testing before it got bogged down in Iraq. Here are the replies of the two presidential candidates in September 2000 to questions in *Arms Control Today*, the publication of the Arms Control Association:

ACT: What is your position on the ratification of the Comprehensive Test Ban Treaty (CTBT)? If the treaty is not ratified, should the United States continue the current testing moratorium?

Bush: Our nation should continue its moratorium on testing. But in the hard work of halting proliferation, the Comprehensive Test Ban Treaty is not the answer. The CTBT does not stop proliferation, especially to renegade regimes. It is not verifiable. It is not enforceable. And it would stop us from ensuring the safety and reliability of our nation's deterrent, should the need arise. On these crucial matters, it offers only words and false hopes and high intentions with no guarantees whatever. We can fight the spread of nuclear weapons, but we cannot wish them away with unwise treaties.

Gore: I believe the Senate rejection of the Comprehensive Test Ban Treaty last year was an act of massive irresponsibility damaging to the security interests of the United States, and if elected president, I will immediately revive the ratification process and seek to rally the full force of American public opinion behind it.

Bush, of course, became president in January 2001 after a decision in his favor by the U.S. Supreme Court about the counting of votes in Florida.

U.S. NATIONAL ACADEMIES CTBT REPORT OF 2012

Interest in the Comprehensive Nuclear Test Ban Treaty (CTBT) resumed when Barak Obama became president in January 2009. John Holdren, a Harvard professor and director of the Woods Hole Research Center, became Obama's science adviser and head of the White House Office of Science and Technology Policy. Later in 2009, Holdren's office and the office of the U.S. Vice President requested that the National Academies conduct a follow-up CTBT study. Its report, published in 2012, was released in both unclassified and classified forms. While CTBT critics continued to find fault with findings in the 2002 report, several critics were generally more complimentary about the 2012 report.

As in the 2002 study, the 2012 report addressed the nuclear weapons stockpile, nuclear monitoring, and potential technical advances to nuclear weapons capabilities that might be gained by other countries from testing that might escape detection. Administration officials also requested views on the utility of on-site inspections as a verification tool and possible effects of undetectable cheating.

Ellen Williams chaired the 2009–2012 CTBT study. Her move from the University of Maryland to BP in London in early 2011 presented some communication complications. Except for two of us, the selected committee members had expertise in nuclear weapons but not nuclear monitoring. In addition to me, the other exception was Theodore Bowyer of Northwest Pacific Laboratories, an expert on the detection of bomb-produced xenon and other radioactive materials.

At first I was the only seismologist on the parent committee, but it would have been too difficult for me to evaluate seismic and other monitoring by myself. Hence, after our first meeting, the main committee established a Subcommittee on Seismology, which I was asked to chair. It was to provide input on detecting, locating, and identifying underground nuclear explosions and determining their yields. All nuclear tests since 1980 had been conducted underground. The Limited Test Ban Treaty of 1963 had banned tests in the atmosphere, space, and underwater but not those underground.

I suggested the names of four other experts in seismic monitoring. The parent committee approved these four as members of the subcommittee: Hans Hartse of Los Alamos; Paul Richards of Columbia University; William Walter of Livermore; and Gregory van der Vink, who had headed the OTA studies on Seismic Verification in 1988 and Containment of Underground Nuclear Explosions in 1989. Our subcommittee analyzed the monitoring of a large number of countries and the main remaining method of possible evasion, decoupling, and wrote extensive sections of the 2012 report on them.

MAJOR CONCLUSIONS OF THE 2012 REPORT

The major conclusions of the 2012 report were:

1. The United States has the technical capability to maintain a safe, secure, and reliable stockpile of nuclear weapons into the foreseeable future

without nuclear explosion testing, provided that sufficient resources for stockpile stewardship are in place. The Stockpile Stewardship Program (which was set up to increase understanding of nuclear device performance and the aging of weapons materials and components) has been more successful than was anticipated in 1999.

2. Seismic and radionuclide monitoring improved substantially during the decade after the 2002 report. Most of the seismic stations of the IMS are now operating and certified, determining data quality, calibration, and integrity as to tampering. U.S. National Technical Means provide additional monitoring capability.
3. Russia and China are unlikely to be able to deploy new types of strategic nuclear weapons that fall outside the design range of their nuclear explosion test experience without several multi-kiloton tests to be confident of their performance. Such multi-kiloton tests would be detectable even with evasive measures. Other countries intent on acquiring and deploying modern, two-stage thermonuclear weapons would not be able to be confident in their performance without multi-kiloton testing as well. Such tests likely would be detectable (even with evasion measures) by appropriately resourced U.S. National Technical Means and a completed IMS network.

STOCKPILE SECURITY

Many people had argued that corrosion, aging, and the phase stability of plutonium would result in short lifetimes for the plutonium pits, which constitute the first stage of thermonuclear weapons. Those pits consist of a hollow shell of plutonium clad in a corrosion-resistant metal, which is surrounded by chemical explosives. When a weapon is detonated, the explosives compress the pit into a supercritical mass and a fission chain reaction occurs. Plutonium pits are a main nuclear component of modern lightweight fission weapons. The behavior of plutonium at or near design yields is critical to the functioning of modern weapons. Work done by the weapons laboratories and reviewed by the JASON group in 2007 indicated long pit lifetimes of eighty-five to a hundred years, much longer than had been assessed earlier. Thus, ascertaining longer pit lifetimes has been a major advance in stockpile reliability.

Life extension programs (LEPs) to repair or replace components and to ascertain the aging of materials were completed as of 2014 for two of the weapons in the U.S. arsenal without the need for nuclear explosion tests. Another LEP is underway as of mid-2017 for the remaining versions of the B61 bomb.

MONITORING

The number of certified stations of the International Monitoring System grew from three in October 2000 to 283 as of mid-2017. The CTBT Organization provides data from areas that the United States previously had difficulty accessing. It also furnishes a common baseline of data to the world's scientific community.

The yield of the North Korean test of 2006 was somewhat smaller than one kiloton, one of many indications that seismic monitoring techniques have improved significantly since the 2002 report.

The classified version of the 2012 report contains a separate section on U.S. verification capabilities, including those of the Air Force Technical Applications Center (AFTAC), which operates the U.S. classified program. While the IMS focuses on global monitoring, the United States also pays great attention to countries of special concern to it.

As discussed earlier, the most significant improvement in radionuclide monitoring since 2002 was the development of very sensitive detectors for radioactive bomb-produced gases such as xenon and argon. They are two of the six inert noble gases, which are difficult to contain following a nuclear explosion. The IMS radionuclide network has gone from being essentially nonexistent in 2002 to a nearly fully functional and robust network with new technology that has surpassed most expectations. The 2012 report states that in at least 50 percent of underground nuclear explosions near one kiloton or larger, even those carried out by experienced testers, xenon may be detectable offsite.

Xenon gases were detected for two of the North Korean underground nuclear tests, including the smallest in 2006. Many past underground

Soviet explosions at Novaya Zemlya are known to have leaked radioactive noble gases. Of course, a nuclear explosion in the atmosphere would produce a greater variety of radioactive products and in larger amounts than an underground test.

WHAT OTHER COUNTRIES MIGHT ACCOMPLISH IN WEAPONS DESIGN BY TESTING AT VARIOUS YIELDS

Table 16.1, from the 2012 report, distinguishes technical achievements that might be accomplished for six different ranges of yields (left column) by countries with no or little prior nuclear test experience (center column) and those with greater experience, such as Russia, China, and the United States (right column). Those with greater experience obviously can accomplish more by testing at a given yield. Nevertheless, they have less to learn because they already have tested many devices and weapons with a variety of yields. It is clear that less can be accomplished as yields become smaller.

The conclusion of the 2012 report indicates that while much can be accomplished by various nations testing at yields greater than 10 kilotons (bottom row), these explosions can be readily detected. Smaller explosions between 1 and 10 kilotons are also unlikely to be concealed. Countries intent on acquiring and deploying modern, two-stage thermonuclear weapons would not be able to have confidence in their performance without multi-kiloton testing.

Note in Table 16.1 that the development of low-yield lightweight boosted fission weapons is possible only for yields of 1 to 10 kilotons for countries with little or no testing experience.

Table 16.1 indicates that proof tests of compact weapons with yields up to 1 kiloton are possible with tests of 0.1 to 1 kiloton by the three most experienced nuclear states. That yield range is likely to be detected in the absence of evasion. Evasive testing via the decoupling scenario is strongly dependent upon the location of an explosion. The Russian and Chinese test sites are well monitored down to very small levels, and neither country contains enough salt at those sites for a significant decoupled test in a large underground cavity.

TABLE 16.1 Purposes and Plausible Technical Achievements for Underground Testing at Various Yields

YIELD (TONS OF TNT EQUIVALENT)	COUNTRIES OF LESSER PRIOR NUCLEAR EXPLOSION TEST EXPERIENCE AND/OR DESIGN SOPHISTICATION (ADVANCES ACHIEVABLE IN THE SPECIFIED YIELD RANGES ALSO INCLUDE ALL OF THOSE ACHIEVABLE AT LOWER YIELDS)	COUNTRIES OF GREATER PRIOR NUCLEAR EXPLOSION TEST EXPERIENCE AND/OR DESIGN SOPHISTICATION (ITEMS IN COLUMN TO LEFT, PLUS)
Subcritical experiments (permissible under the CTBT)	• Equation-of-state studies • High-explosive lens tests for implosion weapons • Development and certification of simple, bulky, relatively inefficient unboosted fission weapons (e.g., gun-type weapon)	• Limited insights relevant to designs for boosted fission weapons
< 1 ton (likely to remain undetected)	• Building experience and confidence with weapons physics experiments	• One-point safety tests • Validation of some unboosted fission weapon designs • Address some stockpile and design code issues
1–100 tons (may not be detectable, but strongly location dependent without evasion)	• One-point safety tests • Pursue unboosted designs	• Develop low-yield weapons (validation of some unboosted fission weapon designs with yield well below a kiloton) • Possible overrun range for one-point safety tests

100 tons–1 kiloton (likely to be detected without evasion; reduced probability of detection with evasion, but strong location dependence)	• Pursue improved implosion weapon designs	• Proof tests of compact weapons with yield up to 1 kiloton
	• Gain confidence in certain small nuclear designs	• Validate some untested implosion weapon designs • Assess stockpile issues and validate some design codes
1–10 kilotons (unlikely to be concealable)	• Begin development of low-yield boosted fission weapons	• Development of low-yield boosted fission weapons
	• Eventual development and full testing of some implosion weapons and low-yield thermonuclear weapons	• Development and full testing of some implosion weapons and low-yield thermonuclear weapons
	• Eventual proof tests of fission weapons with yield up to 10 kilotons	• Proof tests of fission weapons with yield up to 10 kilotons
> 10 kilotons (not concealable)	• Eventual development and full testing of boosted fission weapons and thermonuclear weapons or higher-yield unboosted fission weapons	• Development and full testing of new configurations of boosted fission weapons and thermonuclear weapons • Pursue advanced strategic weapons concepts (e.g., EMP)

Source: National Academies Report of 2012, Table 4-3.

Subcritical experiments called hydrodynamic tests, which involve no nuclear yield, are permitted under the CTBT (first row of Table 16.1). They can be used to test the properties of plutonium at very high shock pressures. The United States and Russia each stated that experiments conducted after 1996 were of zero yield and hence permitted under the terms of the CTBT. Each country observed the preparation and conduct of those experiments by the other in Nevada and Novaya Zemlya, probably by satellite imagery and other national technical means. No seismic waves were detected close to those times from the announced experiments at Novaya Zemlya.

Tests that involve a very tiny nuclear release equivalent to a few pounds (1 kg) of chemical explosive—called hydronuclear—are not permitted under the treaty. During the moratorium on testing in the early 1960s, the United States detonated hydrodynamic explosions in Nevada. It wanted to make sure that its weapons were "one-point safe" —that an accidental blow would generate at most a tiny nuclear yield. Those tests demonstrated that they were one-point safe. Presumably the other major nuclear powers ascertained before they signed the CTBT that their weapons were one-point safe as well. It is unclear if India, Pakistan, and North Korea, which did not sign the CTBT, have weapons that are one-point safe.

In 1994 Kathleen Bailey, a defense analyst, wrote an open technical report titled "Hydronuclear Experiments: Why They Are Not a Proliferation Danger." I have included quotes from her paper here because they indicate that neither hydronuclear nor hydrodynamic experiments can be used to develop advanced nuclear weapons. (In 2014 Bailey and her husband Robert Barker, a former weapons scientist, who have long opposed limitations on nuclear testing, stated that the United States should unsign and renounce its participation in the CTBT.)

Bailey's report was part of the debate in the United States about what yields should be banned under the CTBT as it was being negotiated. She stated, "If a country pursues boosted fission weapons or thermonuclear secondaries [in which the energy from a primary fission reaction compresses and ignites a secondary nuclear fusion reaction], HNEs [hydronuclear explosions] will be an ineffective tool." She went on to say, "Also, HNEs cannot be used to optimize the advanced designs used by existing nuclear weapons states; they are far too low in energy to confirm that boosting will occur in boosted designs, much less provide useful information for staged thermonuclear weapons. They cannot accurately project the neutron multiplication rate at the time a device explodes."

Some CTBT critics claim that Russia continues to perform hydronuclear tests and, if so, that gives them a large advantage. Bailey's 1994 article indicates, however, how little can be learned from those very low yield tests. Can any country develop advanced, lightweight nuclear weapons by such tiny tests and be confident that they will work at yields of say 10 or 100 kilotons? The answer is no.

COMMENTS AT A FORUM HELD BY THE HERITAGE FOUNDATION

Soon after the 2012 NAS report was published, the Heritage Foundation, a conservative research think tank based in Washington, DC, released a transcript of an open forum "Comprehensive Test Ban Treaty: Questions and Challenges." Paul Robinson, a former director of the Sandia weapons lab; John Foster, a former head of Livermore; and Thomas Scheber, a former director of strategic policy in the Department of Defense each spoke and took questions. The quotes that follow are from that transcript.

Robinson said, "Now, whereas one could not accuse that first [2002] report of being an intellectually deep or well-balanced study, I believe you can say that about this report [2012]. It is very much improved, they cover a much larger set of issues, and on some of them they do a very good job. I still have some I find fault with, as you will see. But, it does make far more interesting reading in its thoroughness, and, indeed, I'm proud to say that this group took up the primary issue of this treaty as being foremost about the defense of the country and our national security, and it tried to keep that uppermost. . . . I believe they [the 2012 NAS committee] have done an honest job."

Each of the three, however, had specific criticisms of the 2012 report. It should be remembered that most experts are familiar either with weapons or with verification. The expertise of Robinson, Foster, and Scheber relate to weapons, not verification. They make a number of comments about verification that are decades behind current capabilities.

Robinson said, "One curious weakness in judgment that I'll point out here is the extensive discussions in the report, based on the assumption that if a nation wanted to clandestinely carry out evasive tests, it would choose to do so within its nuclear test site. Now, this is exactly the opposite of what our intelligence community believes; an evader would never attempt to go to the area that we're most heavily monitoring to carry out such an explosion, but certainly countries with large territorial masses would likely find very remote areas in which to conduct their tests, not only because of the ability of great secrecy there, but because they're the farthest from any U.S. monitoring systems."

The 2012 report, in fact, did not state that any country would conduct nuclear explosions only at an existing test site. What is different today from 2002 is that data are now available in near real time from numerous seismic stations in Russia, China, Mongolia, and the now independent countries of Central Asia. For sixty years the United States had sought greater monitoring capabilities in those areas. Their use greatly improves verification throughout various countries of concern to the United States. If stations in one country were unplugged at the time of a suspect event, it would be even more suspect. From his statements, Robinson clearly is not familiar with the present capabilities and how much verification has improved for countries of concern to the United States.

Foster stated, "If we look at the National Research Council report, we see that it talks a lot about detection. Detection is quite different from verification." That statement is no longer correct. Previously, many people thought that to identify an event it needed to be about three times larger than that needed for detection. That criterion was correct when data were only available at large (teleseismic) distances. Once regional seismic data became available, it was no longer correct.

A previous standard rule was that data from at least four seismic stations were needed for detection and the determination of a good location.

Today, good data from one or two regional stations can be used. An example of this is an earthquake of magnitude 2.8 that occurred near the Russian test site at Novaya Zemlya on June 26, 2007. High-frequency P and S waves at the seismic arrays in northern Norway and Spitsbergen were sufficient to identify the event as an earthquake. To obtain a better location, recordings of the 2007 event were cross-correlated with those of previous larger nuclear explosions. Had it been a fully coupled nuclear explosion, its yield would have been about 0.04 kilotons, or 40 tons.

Robinson said, "[W]e in the U.S. labs requested that the permitted test level [under a CTBT] should be set to a level which is in fact lower than a one-kiloton limit, which would have allowed us to carry out some very important experiments, in our view, to determine whether the first stage of multiple stage devices was indeed operating successfully. . . . [T]oday others may be carrying out such experiments without detection, while the U.S. is forbidden to do so."

Debate raged in the United States for nearly twenty years about the determination of Soviet yields near the 150-kiloton limit of the Threshold Treaty. Setting a threshold under the CTBT near the lower limit of detection would be a mistake. Picking a zero nuclear yield was a better choice because debate likely would have gone on for decades about the yields near the lower limit of a very low threshold treaty.

Robinson repeated an old saw: "[T]he 'Little Boy' device, which was first exploded over Hiroshima, had never been previously tested." He did not state that the bomb was extremely heavy and not appropriate for most delivery systems, especially those of Egypt, India, Iran, North Korea, and Pakistan.

Robinson also stated, "Lastly, the CTBT for the first time takes a major step in drastically changing diplomacy of security treaties by surrendering major strategic defense decisions to an international body—a subgroup of the United Nations." In fact, the treaty prohibits the Comprehensive Test Ban Treaty Organization (CTBTO) from reaching conclusions about the nature of a detected event. The CTBTO merely provides for the collection and distribution of data; the United States makes its own determinations. The experience with the five tests by North Korea from 2006 to 2016 indicates that many countries concluded within hours that each of those events was a nuclear explosion.

FINAL WORDS ON DECOUPLED NUCLEAR TESTING: THE VERIFICATION GAUNTLET

The CTBT report of 2012 states:

Evaluation of the cavity-decoupling scenario as the basis for a militarily significant nuclear test program therefore raises a number of different technical issues for a country considering an evasive test:

1. **Is there access to a region with appropriate geology for cavity construction?**
   - Is that geological medium nearly homogeneous on a scale of hundreds of meters [yards]?
   - Can cavities of suitable size, shape, depth and strength be constructed clandestinely in the chosen region?

2. **For a cavity in salt formed by solution mining:**
   - Is enough water available?
   - Can it be pumped out and the brine disposed clandestinely—eight times the cavity volume, plus the final brine fill?
   - How should the very limited experience with conducting decoupled nuclear explosions in salt be taken into account?
   - Can decoupling factors as high as 70 be attained for yields much larger than sub-kiloton (i.e. larger than the 1966 Sterling test)?
   - Can the layered properties of rock sequences for bedded salt be dealt with?

3. **For a cavity in hard rock:**
   - Can mined rock be disposed of clandestinely?
   - Can a country afford the price of mining a large cavity in hard rock?
   - Can uncertainties in rock properties and in orientations and magnitudes of principal stresses be dealt with?
   - Can the presence of joints and faults be detected and dealt with?
   - Can flow of water into the cavity—in either hard rock or salt—be dealt with?
   - Can cavities that depart significantly from a spherical shape be used?
   - Should a decoupling factor no larger than 10 to 20 be assumed?

4. **Can collapse of a cavity during construction and in a decoupled test be avoided?**
    - Can surface deformation potentially detectable by interferometric synthetic aperture radar (InSAR) both during and after construction and following the test be minimized?

5. **Can radionuclides be fully contained from a decoupled explosion?**
    - Take into account that noble gases can be detected today at much smaller concentrations than a decade ago.
    - Take into account that radionuclides have leaked from many previous nuclear explosions in hard rock at Novaya Zemlya and Eastern Kazakhstan and the few in granite at the Nevada Test Site.

6. **Can the site be chosen to avoid seismic detection and identification, given the detection thresholds of modern monitoring networks and their capability to record high frequency regional signals?**
    - Can the limited practical experience with nuclear tests in salt, and very low-yield chemical explosions in hard rock, be extrapolated to predict the signals associated with nuclear testing in cavities in hard rock?
    - Can the size of a test be made small enough to deal with future advances in detection and identification capabilities?

7. **Is there such a region that is suitably remote and controllable, and that can handle the logistics of secret nuclear weapons testing?**
    - Can secrecy be successfully imposed on all of the people involved in the crosscutting technologies of a clandestine test program, and on all who need to know of its technical results?
    - Can the tester avoid compromising security by conducting a nuclear test in a region containing a hostile ethnic group or a civil war? Can the test be conducted outside one's own territory?

8. **Can nuclear explosions of large enough yield be carried out secretly, and repeated as necessary, to support the development of a deployable weapon?**
    - Can those carrying out a decoupled test be sure that the yield will not be larger than planned, and thus only partially decoupled?

- Can a minimum of drill holes, cables, and specialized equipment be used and yet obtain necessary information about the characteristics of nuclear device(s)?
- Can the site be cleaned up before an on-site inspection team arrives?

9. **Can a clandestine test in a mining area be hidden in one of a series of ongoing large chemical explosions?**
   - Can suitable rock for a decoupled test be found below coal, other minerals, and sedimentary rock in which large chemical explosions are used in mining?

This verification gauntlet faces a country wanting to conduct a decoupled nuclear explosion with confidence it would not be detected and identified. As stated earlier, Russia and China are unlikely to be able to deploy new types of strategic nuclear weapons that fall outside the design range of their nuclear explosion test experience without several multi-kiloton tests to build confidence in their performance. Each already conducted tactical nuclear explosions prior to signing the CTBT.

The 2012 report stated that yields of 0.1 to 1 kilotons are likely to be detected without evasion and at reduced probability with evasion (but with a strong location dependence). The probabilities of nuclear explosions' being detected and identified are even better for decoupled explosions at the test sites of Russia at Novaya Zemlya, China at Lop Nor, North Korea, and India. Because salt is not present in sufficient thicknesses at any of those sites, decoupled tests would have to be conducted in hard rock.

Weapons designers would want to collect a variety of data from a clandestine decoupled nuclear explosion in addition to merely knowing that it detonated and generated an approximate yield. This goes against having only a sparse set of recording instruments. Before 1996 the United States used huge amounts of instrumentation and large cranes to emplace nuclear devices underground (figure 16.1).

High-level administrators of the weapons labs who were members of the two Academies studies were not proponents of decoupled testing. The 2012 report found that yields of nuclear explosions that are fully coupled

FIGURE 16.1

Signal cables and test device being lowered down a test hole in Nevada. Photo credit: Department of Energy.

Source: Office of Technology Assessment, 1988.

and that might not be detectable are very small. Russian scientists stated to me that the partially decoupled explosion at Azgir in Kazakhstan in 1976 was their only nuclear test utilizing decoupling. I know of no other Russian decoupled nuclear test.

17

STRATEGIC NUCLEAR WEAPONS: SOVIET AND U.S. PARITY

In 1987 Daniel Davis of Stony Brook University and I published "The Yields of Soviet Strategic Weapons" in *Scientific American*. We tried to ascertain if the Soviet Union possessed more powerful nuclear weapons and more capable intercontinental-range (strategic) delivery systems than the United States. To do so, we determined yields from seismic waves generated by large Soviet explosions and combined them with published information on so-called throw weights—how many tons (or kilograms) of nuclear warheads their missiles could carry. We found the two countries were at near parity in intercontinental delivery systems and the nuclear weapons they carried.

We matched the occurrences of various large nuclear explosions with published information on the dates the delivery systems were first tested and then deployed. We used calculations of yields of Soviet explosions at Novaya Zemlya, their arctic test site, that I had obtained with Graham Wiggins, a student who worked for me the summer before, using a combination of seismic body and surface waves.

The *Scientific American* article was motivated by reports in the media of hugely inflated estimates of the capabilities of Soviet missiles and the yields of each of the multiple nuclear weapons they carried. Senator Robert Dole, a high-ranking Republican, announced that the yield of *each* of the eight to ten nuclear warheads carried by the large Soviet missile, the SS-18, was fourteen times larger than what we subsequently concluded in 1987. Dole's huge overestimates had a great impact on U.S. defense policy and fueled fears about the consequences of a Soviet nuclear attack.

For each of the Soviet intercontinental land-based ballistic missiles (ICBM) known by NATO as the SS-17, SS-18, and SS-19, two major versions (mods) existed—one with a single very large warhead and a second with multiple, independently targetable reentry vehicles (MIRVs). The capabilities of the very large SS-18 missile understandably were of great concern to the United States. At the negotiations for the Threshold Test Ban Treaty (TTBT) in June 1974, General Edward Rowny, who represented the U.S. Joint Chiefs of Staff, had focused on the large payload of nuclear weapons of the SS-18 and not much on the testing of yields of warheads the United States should seek to prohibit under the TTBT.

If the United States had insisted in 1974 on a TTBT with a lower yield limit of, say, 50 kilotons, tests as large at 150 kilotons would not have been permitted after March 31, 1976, and the USSR might well not have been able to fully test multiple warheads for its SS-17, SS-18, and SS-19 strategic missiles. At that time, the United States was ahead of the USSR in testing and deploying land- and sea-based missiles with multiple warheads. The TTBT allowed very large tests to be conducted for nineteen months after it was signed, which gave the Soviet Union time to catch up with the United States in testing multiple warheads.

Paul Nitze, who served as a U.S. high-level official and as a special adviser to the president on arms control, testified before Congress in July 1979 about the SS-18 and other Soviet long-range delivery systems. His estimate of 750 kilotons for each of the multiple warheads on the SS-18, as published in the *Congressional Record—Senate*, was much smaller than Senator Dole's. Nitze expressed great fear that the Soviet Union would obtain a substantial lead over the United States in the nuclear arms race with its larger missiles and their many warheads of high yield.

Interestingly, on May 31, 1979, journalist Walter Pincus wrote a front-page story in the *Washington Post* stating that the U.S. intelligence community had substantially downgraded its estimate of the explosive yields of each of the eight to ten weapons carried by the SS-18, from the previous year's 1200 kilotons to 600 kilotons in 1979. He reported that analysts had successively lowered their previous estimates of the yield of each of its warheads from 3000 to 2000 to 1200 and thence to 600 kilotons. If correct, this represented a reduction in yield by a factor of five and was close to the yield Davis and I published in 1987. If, as originally estimated, the SS-18 missile carried ten warheads of 3000 kilotons each, its total payload would have been 30,000 kilotons. If it carried eight warheads of 600 kilotons each, the total yield for each SS-18 would have been 4800 kilotons, a huge difference. In a sense, this was a continuation of the "yield wars" described in chapter 10.

Pincus stated, "This sharp reassessment of the SS-18's yield could play a major role in the coming Senate debate on the U.S.-Soviet strategic arms limitation treaty (SALT II), according to sources on Capitol Hill." President Jimmy Carter and Soviet Premier Leonid Brezhnev signed that bilateral treaty on June 18, 1979, but the U.S. Senate failed to ratify it. Debate over the capabilities of Soviet missiles and their warheads, along with the Soviet invasion of Afghanistan, contributed to the Senate's inaction. Nevertheless, the two countries honored the terms of SALT II until the Reagan administration withdrew from the treaty in 1986 after accusing the USSR of violating it.

Pincus said that the six warheads on the Soviet's second largest missile, the SS-19, in 1979 were each rated as 500 kilotons instead of the previously estimate of 800 kilotons. Davis and I also obtained 500 kilotons for the yield of each of its multiple warheads.

It is unfortunate the United States focused so much effort on the yields of tests and alleged Soviet cheating at Eastern Kazakhstan rather than on the much larger nuclear explosions that were set off at Novaya Zemlya. The yields of weapons deployed by the Soviet Union in the 1970s on its strategic delivery vehicles were large enough that full-yield tests had to be conducted at Novaya Zemlya between 1966 and 1975. It is for that reason that Davis and I decided to work on those tests and their implications for the capabilities of Russia's strategic weapons.

The United States was not permitted to monitor any of the nuclear explosions on Novaya Zemlya, nor was it allowed to conduct a Joint Verification Experiment there. Unlike Eastern Kazakhstan and sites of several peaceful nuclear explosions, the Russians did not publish exact yields for any of their explosions at Novaya Zemlya until twenty-two years after they signed the Threshold Test Ban Treaty.

In 1996 and 2000, the ministries of Atomic Energy and Defense of the Russian Federation finally published some information on the numbers and yields of nuclear tests. Many of the tests at Eastern Kazakhstan, as well as peaceful explosions, consisted of single explosions. Exact yields published by the Russian Federation for several of those explosions agree with estimates made by Western scientists, including me, and in 1992 by Ringdal, Marshall, and Alewine.

Nevertheless, the Russian information is more difficult to interpret for Novaya Zemlya explosions. It indicated that many large tests between 1966 and 1976 at the northern site consisted of multiple explosions that were detonated very close in time, in what are called salvos. In contrast, they listed three of the four explosions at the southern Novaya Zemlya test site from 1973 to 1975 as single events.

Why did the Soviet Union test two or more nuclear devices in a salvo? The northern site at Novaya Zemlya is located farther north than the northernmost part of Alaska. Working and testing in the far arctic at Novaya Zemlya were limited to times of less severe weather, ice conditions, and quite possibly by the amount of sunlight. Most of those underground tests occurred during September and October, with only a handful in August, November, and December. Hence, the Soviets had little time each year to conduct large explosions at Novaya Zemlya, especially in the period between the signing of the TTBT in July 1974 and its start date in early 1976.

Russian publications in 1996 and 2000 listed the sum of the yields tested per year at Novaya Zemlya but gave only a broad range of yields for underground explosions on a particular date. For example, on August 23, 1975, they listed a salvo consisting of eight explosions—four with yields of 150 to 1500 kilotons, two of 20 to 150 kilotons, and two of 0.001 to 20 kilotons. Thus, the individual yields of the four largest explosions in that salvo are very uncertain.

Many people in the United States, including Davis and me, were not aware in 1986 that large numbers of explosions at the northern test site at Novaya Zemlya were fired at nearly the same time. Subsequently, U.S. and British scientists identified several other events that consisted of more than single explosions.

Since the Russian government published the cumulative yield per year for Novaya Zemlya, one of the better comparisons with their data is the summed yields we published. The sum of the yields in the Russian lists for the two test sites on Novaya Zemlya between 1966 and 1975 is about 1.7 times greater than what we published. Since many of the salvos consisted of two or more individual large explosions, our published yields of weapons on Soviet delivery systems probably were not as uncertain as a factor of 1.7.

Why did we do poorly for yields at Novaya Zemlya but so well for Eastern Kazakhstan and Soviet peaceful explosions? We made two incorrect assumptions when we calculated Soviet yields at Novaya Zemlya in 1987. One was that seismic waves passing through the upper mantle beneath those large tests did so in the same way as beneath Eastern Kazakhstan. The other was that events on a given date were single explosions.

I now believe that our calculated yields for salvos at the northern site on Novaya Zemlya were too small. Most explosions at the two arctic sites occurred at northern Novaya Zemlya. Our magnitude corrections (m_b bias) with respect to Nevada were like those we calculated for Eastern Kazakhstan—that is, about 0.35 magnitude units. We applied that correction in calculating yields for the two Novaya Zemlya sites using explosions of known yield and m_b in Nevada and the Aleutians.

The Russian lists of 1996 and 2000 stated that their largest underground nuclear explosion occurred on September 12, 1973, at the northern Novaya Zemlya site. Davis and I had concluded, however, that it was the second largest Soviet explosion and that one at southern Novaya Zemlya on October 27, 1973, was larger. Two Russian references give the yield of the single great event of 1973 at southern Novaya Zemlya as 3500 kilotons. Subtracting that yield and that of a much smaller event from the Russian sum of 7820 kilotons for 1973 gives a yield of about 4200 kilotons for the largest Soviet underground test on September 12, 1973, at the northern site. Some Russian experts told my colleagues that event was smaller, around 3500 kilotons. In any case, we clearly underestimated its yield at 1830 kilotons.

In 1987 Davis and I calculated an average yield of 3450 kilotons for the single explosion at southern Novaya Zemlya in October 1973, which was nearly identical to the published Russian yield. Thus, I conclude that the magnitude correction we used was about right, 0.35 m_b units, for southern Novaya Zemlya but was too large for the northern site. The explosion on September 12, 1973, was not the largest Soviet underground test as they claim unless the magnitude correction or bias for the northern site on Novaya Zemlya is smaller than that for the southern site.

In 1979 Marshall and colleagues used the speed of P waves at the top of the Earth's mantle as an indirect indication of the magnitude corrections to be made for various test sites. They concluded that northern Novaya Zemlya was more attenuating than Eastern Kazakhstan. I now concur with their judgment.

Davis and I estimated the yields of the single very large warheads carried by the SS-17 and SS-19 missiles as 2000 and 3500 kilotons based on our calculated yields for the two largest underground Soviet tests of September 12 and October 27, 1973. My revised estimates for those two are about 3500 and 4200 kilotons. The larger yield is comparable to that of the largest U.S. underground test, *Cannikin*, which was intended for the U.S. Spartan anti-ballistic missile. The *Cannikin* device, however, was not deployed as a weapon, whereas the Soviet nuclear devices almost certainly were.

On April 5, 1985, Jack Anderson and Dale van Atta reported in the *Washington Post* that a warhead gap existed between the number of Soviet warheads estimated by the CIA, 6500, and by the U.S. Defense Intelligence Agency (DIA), 8500. They stated, "But at the last minute, the DIA chickened out and allowed publication of the CIA's less scary estimate." They went on to say, "A Soviet SS-18 could hold two dozen warheads. On that both agencies agree." They then concluded, "In this case, we'd be inclined to lean toward the DIA estimate." This is the only instance I know of in which the SS-18 has been credited with carrying more than eight to ten warheads.

Jack Anderson had published incorrect estimates of Soviet yields in the *Washington Post* in August 1982, stating, "the Soviets appear to have exceeded the 150-kiloton limit [of the Threshold Treaty] at least 11 times since 1978. One test in September 1980 was clocked at a likely size of 350 kilotons, according to my sources." We now know that his estimates of those eleven yields at Eastern Kazakhstan were too large. I think Anderson and van Atta most likely gave inflated estimates for the SS-18.

In summary, I think the better estimates are that the yield of the individual (MIRVed) warheads on the Russian SS-18 intercontinental missile is about 500 to 600 kilotons and the yield of those on the SS-19 is about 500 kilotons. Both of those missiles were still deployed as of mid-2017. While Russia has more long-range warheads deployed on land-based missiles, the United States has more of its assets at sea. The published yield of the W-88 warhead on U.S. Trident II submarines is 475 kilotons.

STRATEGIC DELIVERY SYSTEMS AND NUCLEAR WEAPONS OF THE UNITED STATES AND RUSSIA

The number of long-range nuclear weapons maintained on high alert by the Russian Republic and the United States has been reduced through various bilateral treaties since our paper in 1987 (figure 17.1). Nevertheless, Russia and the United States each still have many more nuclear weapons than other countries.

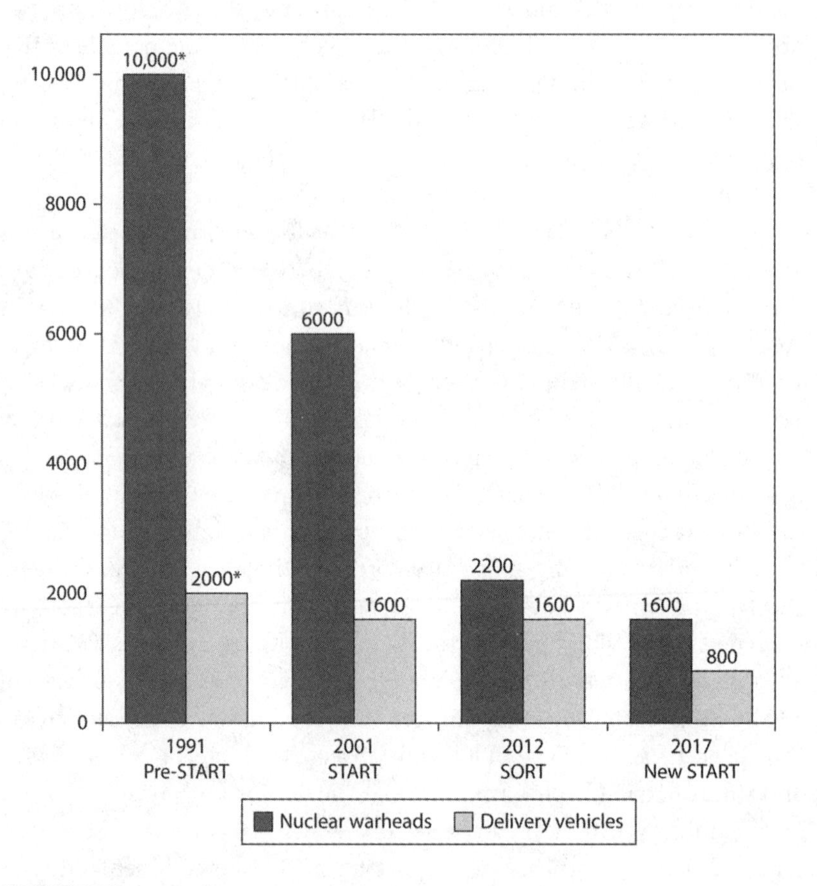

FIGURE 17.1

Numbers of long-range (strategic) nuclear weapons and their delivery vehicles permitted under bilateral treaties between the United States and Russia.

Source: U.S. State Department.

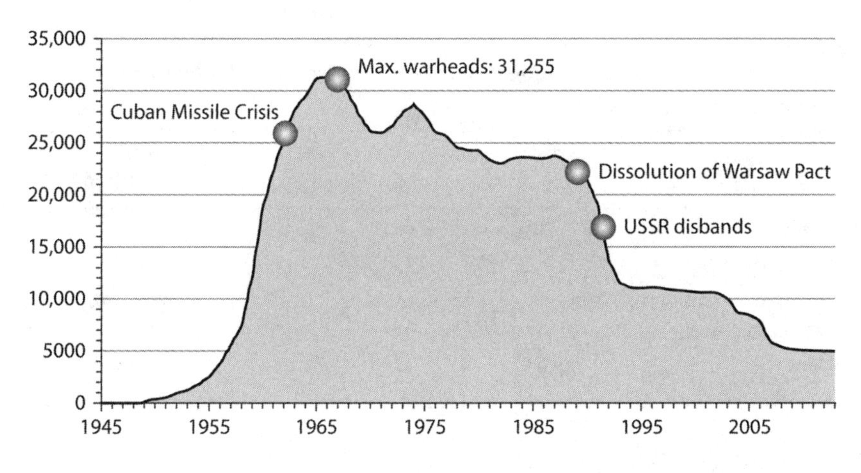

FIGURE 17.2

U.S. nuclear stockpile, 1945–2014.

Source: Redrawn from fact sheet, U.S. Department of State, 2014.

Figure 17.2 indicates that the number of U.S. nuclear warheads reached a peak of about 31,255 in 1967. Table 17.1 lists the numbers of offensive (long-range) nuclear arms of Russia and the United States at the end of 2013. The number of warheads that could be delivered rapidly by the United States and Russia is still huge—1688 and 1400—as are the numbers of facilities, cities, and people that could be targeted. A major exchange of weapons by the two nuclear superpowers could result in the immediate deaths of hundreds of millions of people in the two countries as well as comparable numbers in other countries in the Northern Hemisphere. The longer-term consequences, which are huge, are examined in the last chapter.

Russia and the United States each have a different sense of what constitutes "strategic delivery systems" for nuclear weapons. The United States, separated from Russia, China, and other nuclear states by large oceans, understandably makes a distinction between intercontinental delivery systems and others of intermediate and shorter range. Russia is faced with nuclear states at varying distances. Thus, it does not make as strong a distinction between intercontinental delivery systems and those for possible use within Asia and Europe. Various bilateral treaties between Russia

TABLE 17.1 **Bureau of Arms Control, Verification, and Compliance Fact Sheet, January 1, 2014: Numbers of offensive (strategic) nuclear arms of Russia and the United States**

CATEGORY OF DATA	UNITED STATES OF AMERICA	RUSSIAN FEDERATION
Deployed ICBMs, deployed SLBMs, and deployed heavy bombers	809	473
Warheads on deployed ICBMs, on deployed SLBMs, and nuclear warheads counted for deployed heavy bombers	1688	1400
Deployed and non-deployed launchers of ICBMs, deployed and non-deployed launchers of SLBMs, and deployed and non-deployed heavy bombers	1015	894

ICBMs denote intercontinental land-based ballistic missiles; SLBMs, submarine-launched long-range ballistic missiles. Data are from the biannual exchange of data required by the treaty and are declared current as of September 1, 2013.

Source: U.S. Department of State.

and the United States do specify the distance ranges for various nuclear delivery systems.

India and Pakistan each have about a hundred nuclear weapons, and their numbers are growing. A major exchange between them could well involve several hundred million causalities. While India is Pakistan's sole major adversary, India faces threats from both Pakistan and China, which has been an ally of Pakistan for decades. A major exchange involving the three countries could cause more than half a billion fatalities, and the conflict might spread to other nuclear powers. India and Pakistan may well not have weapons that are as safe as those of the United States, Russia, and China. In addition, their ability to monitor actions by other countries is likely not as great.

The world has been fortunate that nuclear weapons have so far not been used by one state against another since 1945, but this could end at a moment's notice. I do not derive much comfort from comments by some analysts that nuclear weapons have prevented a world war since 1945.

U.S. Secretary of Defense Robert McNamara noted during the Cuban Missile Crisis of 1962 that it did not matter that the United States had many times more warheads than the Soviet Union. If the Russians had delivered just ten warheads on the territory of the United States, it would have been a disaster beyond what we had ever experienced. For that reason, he said, the United States, even in 1962, would not have carried out a preemptive strike against the Russians for fear of what they could do to us in return. The United States and Russia continued to keep their strategic arsenals on a par with one another for decades after 1970.

18

NUCLEAR WAR, FALSE ALARMS, ACCIDENTS, ARMS CONTROL, AND WAYS FORWARD

In this final chapter I take a broad look at a number of issues related to the control of nuclear arms that extend beyond nuclear testing. I examine the consequences of a major nuclear exchange, the concept of nuclear winter, attempts to limit nuclear weapons and their delivery systems, the possibility of eventually eliminating all nuclear weapons, and the role of scientists in reducing the likelihood of a nuclear exchange. My list is only partial.

DANGERS OF HAIR-TRIGGER ALERTS AND UNAUTHORIZED LAUNCHES OF DELIVERY SYSTEMS FOR NUCLEAR WEAPONS

Nearly everyone agrees that a major exchange of nuclear weapons would be a great disaster of unprecedented dimensions and horrors. Nevertheless, preventing such an exchange involves considerable disagreement. I am of the school that thinks that the use of even a few nuclear weapons would be a major catastrophe, and that nuclear weapons have only one use—deterring others from using them. Some people in the United States, however, believe that nuclear weapons could be used to destroy chemical and biological weapons, destroy deep bunkers for nuclear arms, attack terrorists, or conduct a limited war.

It is a mistake to think that chemical, biological, and nuclear weapons —sometimes collectively called weapons of mass destruction—are of equal lethality. While deadly, scenarios that involve chemical and biological weapons are nearly always much less destructive than those involving nuclear weapons.

A decision by the Russian Republic to attack the United States with hundreds to a thousand nuclear weapons—a so-called "bolt out the blue"—has not been considered a likely scenario since the end of the Cold War in 1991. The greater danger today is that many nuclear weapons are on hair-trigger alert and might be used in a rapid response (fire on warning) in response either to a false alarm (believing one is under attack) or to an unauthorized launch of one or more delivery vehicles. Just the firing of one missile with multiple warheads by one country against the other—such as either the Russian SS-18 or the U.S. Trident II—could destroy four to eight major cities. The threat of computer hacking is now an additional concern.

Russian land-based intercontinental ballistic missiles (ICBMs) can reach the United States in as little as twenty-five minutes, submarine-launched missiles in as little as ten minutes. Once the United States detected their launch, even less time would remain before impact. This would allow high-level officials only five to ten minutes to deliberate whether they should launch U.S. strategic forces against Russia before command and control systems and land-based missiles sustained damage. About two minutes would be needed to transmit launch orders, three minutes for firing those missiles, and a few more minutes for them to fly a safe distance from their home bases.

A strong rationale exists for either the United States or Russia to fire on warning, launching a retaliatory strike while enemy nuclear missiles presumably are still en route and before detonations occur, to avoid loss of land-based missiles, bombers, and command and control facilities. Nevertheless, fire on warning is a particularly dangerous practice because it could well occur in response to a false alarm. This is particularly true for land-based missiles with multiple warheads because one or two incoming warheads could destroy a single missile with up to ten warheads. Russia has a larger percentage of its long-range warheads on land-based missiles, many of which contain multiple warheads (MIRVs), while the United States has more of its strategic warheads on submarines. These U.S. assets deployed well at sea are not vulnerable to a sudden attack, but their land-based command and control facilities are.

The United States also has the better early warning systems. Russia, therefore, is more likely to fire on warning in response to a false alarm that is not identified as such in the ten minutes or so required to launch its larger land-based assets. Russia may believe it has even less warning time because it fears the use of missiles launched from U.S. Trident submarines in the northeast Atlantic Ocean off Norway. Once a few nuclear weapons are fired, an orgy in which many more weapons are set off seems highly likely. Missiles, unlike bombers, cannot be recalled or redirected to another site once they are launched.

NUCLEAR ACCIDENTS AND FALSE ALARMS

I have compiled from a variety of sources a list of accidents involving nuclear weapons and false alarms of impending nuclear attack that have occurred during the past sixty-five years. Soviet accidents have likely been underreported.

1. *Northern British Columbia, February 13, 1950:* A U.S. B-36B bomber crashed after jettisoning a nuclear bomb, the first known loss of a nuclear weapon. It did not cause a nuclear explosion.
2. *Goldsboro, North Carolina, January 23, 1961:* Two U.S. hydrogen bombs were accidently dropped when the B-52 carrying them broke up in midair. Three safety mechanisms on one bomb failed; the fourth and final mechanism worked and prevented explosions of several megatons.
3. *Palomares, Spain, January 17, 1966:* A U.S. B-52 carrying four hydrogen bombs collided with a tanker during aerial refueling. Three of those weapons were found on land. The nonnuclear explosives in two of them detonated upon impact with the ground, resulting in radioactive contamination of a 0.78 square mile (2 sq km) area. This is an example of weapons' being "one-point safe" in that strong blows to the chemical explosives did not detonate one or more nuclear explosions. The fourth

weapon fell into the Mediterranean Sea, where it was recovered intact after a two-and-a-half-month search.

4. *Thule, Greenland, January 21, 1968:* A B-52 crashed onto sea ice after a cabin fire. Four hydrogen bombs ruptured and dispersed widespread radioactive contamination. No nuclear explosions occurred.
5. *Use of a wrong tape, November 9, 1979:* Computers of the North American Aerospace Defense Command (NORAD) indicated a full-scale, preemptive Soviet attack against the United States. U.S. military officers feared the worst for six minutes before recognizing it as a false alarm. A NORAD technician had mistakenly loaded a simulation into the system without identifying it as such.
6. *Damascus, Arkansas, September 1980:* A nonnuclear explosion occurred in a Titan II missile silo when an airman dropped a socket from a wrench that pierced the skin of the missile. The warhead landed about 100 feet (30 m) from the entry gate of the launch complex. No radioactive material was dispersed, and a nuclear explosion did not occur.
7. *A Soviet error, September 26, 1983:* The Soviet early warning system mistook bright flashes to mean that the United States had launched five nuclear missiles at the USSR. Disobeying orders, a Soviet lieutenant colonel in charge decided against informing his superiors. He reasoned correctly that a system malfunction had occurred because the United States would not have launched an attack of just five missiles.
8. *A failed communication, January 25, 1995:* A communiqué from the Norwegian government describing the launch of a research rocket to study northern lights never reached the Russian military. For a few minutes, Russian radar operators believed they were under attack by the United States. The alarm reached high levels of the Russian government.
9. *Hacking into U.S. launch orders, 1990s:* An in-depth investigation of safeguards found an electronic "backdoor" to the naval communications network used to transmit launch orders to U.S. Trident missile submarines. Whether the network was actually ever hacked into is not known.
10. *Missing U.S. cruise missiles, August 29–30, 2007:* Six cruise missiles armed with nuclear warheads were loaded onto a U.S. Air Force plane,

flown across the country, and unloaded. For thirty-six hours, no one knew where the warheads were, or even that they were missing.

BALLISTIC MISSILE DEFENSE

Present versions of U.S. defensive missiles, which do not contain nuclear warheads, destroy a target missile by either colliding with it or detonating a chemical explosion close by. It must get very close to its target, which is moving extremely fast, in a very short amount of time. It is like hitting a bullet with a bullet. Defensive missiles are still in the process of being developed and tested. Even though many have already been deployed, many U.S. tests have missed their targets.

Missile defense is not a new concept. Various U.S. versions go back to the 1960s, including the nuclear explosion code-named *Cannikin* of about 4500 kilotons, which was conducted at Amchitka Island in the Aleutians in 1971. It was intended for use with an early missile interceptor, which may have been deployed briefly before Congress defunded it. In 1983 the Reagan administration proposed a Strategic Defense Initiative (SDI), an ambitious project that would have constructed a space-based antimissile system intended to make nuclear weapons "impotent and obsolete." This program was dubbed "Star Wars." One version of SDI involved high-powered lasers on U.S. satellites that could be aimed in specific directions at missiles soon after they were launched. Because the satellites would have been vulnerable to a preemptive attack, the system would have to be fired in a matter of minutes once a real or supposed launch of Russian or other missiles was detected.

In December 2001 President George W. Bush gave notice to Russia that in six months the United States would withdraw from the bilateral Anti-Ballistic Missile Treaty of 1972. The United States did withdraw and went on to deploy defensive missiles in Alaska and Eastern Europe and on ships in the eastern Mediterranean.

Advanced long-range missiles carrying nuclear weapons typically involve several stages. They are most vulnerable in the initial stage after launch, when they move relatively slowly from the Earth's surface through the atmosphere. Nevertheless, once they are detected, little time is available to attack them with anti-ballistic missiles. Now lighter and having burned much fuel, intercontinental and intermediate-range missiles with one or several warheads then travel long distances above the atmosphere through the near vacuum of space. Many decoys or penetration aids can be released from the missile while in space. Lightweight decoys are very difficult to distinguish from warheads in space because neither decoys nor warheads are slowed by atmospheric friction. The warheads finally reenter the atmosphere above their targets, descending very fast. Very little time is available at this point to destroy them.

A major missile attack by either Russia or the United States likely would involve hundreds of warheads. To be successful, a defensive system must destroy all of them, which seems exceedingly remote. Even if only a small percentage of them were not destroyed and reached their targets, the immense destruction would be a national disaster unprecedented in human history. In fact, the United States has long emphasized that its ballistic missile defense system cannot counter more than a small percentage of Russian intercontinental missiles. To keep the momentum up and to fund current anti-ballistic missile systems, the United States has emphasized that they are intended as a counter against "rogue" countries.

The emphasis now is on destroying one or a few nuclear missiles launched by either Iran or North Korea against either the United States or its allies. Deploying anti-missile systems to do that understandably creates fears in China and Russia for the safety of at least some of their existing missiles. Either of those countries is likely to respond by deploying more intercontinental missiles, making the United States less safe, and both are likely to resist reducing the number of intercontinental missiles further in future arms control agreements.

A missile launched by either Iran or North Korea that carried a nuclear warhead anywhere near the United States would almost certainly trigger a massive attack by the United States. In my estimation, the United States is much too concerned about a nuclear attack on either it or its allies by either Iran or North Korea. Israel, a strong U.S. ally, can be counted on to defend itself against attacks by Middle Eastern countries. Israel's nuclear capabilities are large enough to provide sufficient deterrence against Iran's using nuclear weapons against it. The United States should worry more about North Korea and Iran furnishing either nuclear weapons or know-how to other countries and to terrorists.

Many nonscientists in the United States do not seem to understand that most arms control experts in the scientific community regard a successful defense system against tens to hundreds of incoming warheads as a pipe dream. This is an instance in which technical considerations really are critical in formulating U.S. national security policy. In 2014 Israel's "iron dome" shot down roughly half of the crude, short-range rockets that were fired into Israel from Gaza. Success in intercepting *some* of those rockets should not be confused with the immensely more difficult task of distinguishing many warheads that travel though space along with sophisticated decoys. It is important to note that nuclear weapons can also be delivered clandestinely by means other than ballistic missiles, including by trucks and ships.

Russia and China have long feared U.S. technological advances. Hence, each of them may well cooperate with the United States in further reducing nuclear weapons and their delivery systems if serious limitations are placed on anti-missile systems. This will be a tough pill to swallow for many people in the United States who believe in simplistic arguments that ABM systems can protect us from nuclear attack. Defense contractors and politicians who invoke patriotism continue to argue for new ABM systems. Thus far, for domestic political reasons, no U.S. president, no matter how intelligent, has dared to state that ballistic missile defense is unworkable and hugely expensive and that it diverts funds from many other things of great value to the nation, including other aspects of national security. The United States likely could obtain many important concessions from Russia and China in a "grand bargain" that would either eliminate or seriously restrict ballistic missile defense.

In 1986 President Reagan rejected limits on missile defense during his summit meeting with Russian general secretary Gorbachev in Reykjavik, Iceland. Reagan could not understand that his vision of a nuclear-free world was not compatible with his excitement for ballistic missile defense. Gorbachev, while interested in warhead reduction, also wanted major limits on defensive missiles. Placing severe limits on missile defense continues to be a major Russian objective.

As Steven Weinberg, a theoretical physicist at the University of Texas and a Nobel laureate in physics, stated in 2002, "There is nothing more important to American security than to get nuclear forces on both sides down at least to hundreds or even dozens rather than thousands of warheads and especially to get rid of MIRVs, but this is not going to happen if the United States is committed to a national missile defense."

ABOLISHING NUCLEAR WEAPONS

On January 15, 2008, George Shultz, William Perry, Henry Kissinger, and Sam Nunn coauthored "Toward a Nuclear-Free World" in the *Wall Street Journal*. Shultz was U.S. secretary of state from 1982 to 1989; Perry was secretary of defense from 1994 to 1997; Kissinger was secretary of state from 1973 to 1977; Nunn formerly chaired the Senate Armed Services Committee and was very involved with nuclear weapons and arms control. The title of their article was surprising because each of them had been moderate to conservative in their political and arms control views.

They stated, "The accelerating spread of nuclear weapons, nuclear know-how and nuclear materials has brought us to a nuclear tipping point. We face the very real possibility that the deadliest weapons ever invented could fall into dangerous hands. . . . With nuclear weapons more widely available, deterrence is decreasingly effective and increasingly hazardous." They went on to say, "Without the vision of moving toward zero, we will not find the essential cooperation required to stop our downward spiral." They listed many world leaders who supported their views.

Their call to move toward zero brought out many critics who mentioned old, much-used arguments: nuclear weapons cannot be disinvented; some countries will successfully hide weapons; not all nations will comply. One commenter on the proposal by Shultz and colleagues noted, however, that ending the Cold War was more utopian than the elimination of nuclear weapons and that 95 percent of nations are already nuclear free. Of course, it will take time to move toward zero, but many steps can be taken in the meantime. There is a need for good positive thinking about verifying movements toward zero.

Shultz and others advocate a series of steps to reduce the nuclear threat. An obvious one is to increase the decision times for the launch of nuclear-armed ballistic missiles to reduce risks of accidental or unauthorized use.

They also advocate the Comprehensive Nuclear Test Ban Treaty, a stronger nonproliferation regime, and an international system to manage the risks of the nuclear fuel cycle.

I think Schulz and others were weak on ballistic missile defense (BMD). They said, "Undertake negotiation toward cooperative multilateral ballistic-missile defense and early warning systems. This should include agreements on plans for countering missile threats to Europe, Russia and the United States from the Middle East."

They do not mention one measure that I think is very important: a considerable reduction or elimination of missiles with multiple, independently targetable reentry vehicles (MIRVs). Because each of those missiles carries several weapons, they are more likely to be launched in response to warnings than missiles with single warheads. The understandable fear is that missiles with multiple warheads will be destroyed if they are not used quickly.

Kissinger remarked decades ago that the United States should have thought through more carefully the development and deployment of MIRVs. They were (and still are) very destabilizing to arms control. One missile with multiple warheads could destroy three to ten missiles of the other superpower.

I think that Shultz and others should have been stronger about nuclear weapons that are still deployed in Europe. They called only for a dialogue with other NATO countries and Russia about the forward deployment of weapons, a careful accounting of them, and their eventual elimination. Moving to eliminate tactical and other remaining nuclear weapons in Europe, including western Russia, seems to me to be amenable to negotiation. Those weapons are no longer deployed in Belarus, Kazakhstan, or Ukraine—countries that were formerly part of the Soviet Union.

PROPOSED NEW U.S. DELIVERY SYSTEMS

The administration of President Barack Obama stunned arms control advocates by embarking on an aggressive effort to upgrade the military's nuclear weapons programs, including requests to buy twelve new missile-firing submarines, up to a hundred new bombers, and four hundred land-based missiles over the next thirty years. Russia, whose nuclear delivery systems were degraded after the breakup of the Soviet Union, has made recent efforts to reverse that trend under President Vladimir Putin. That reversal, Russia's annexation of the Crimea, threats to Ukraine, a more aggressive China, and warfare with Muslim extremists have contributed to calls in the United States for increased nuclear capabilities. However, massive rebuilding of U.S. nuclear forces appears to have caused Russia to modernize more of its nuclear arsenal than previously planned in an attempt to keep up with the United States. These contribute to a new Cold War mentality and increase the dangers of nuclear war.

NUCLEAR WINTER

Over the past several decades, a number of authors have proposed that the detonation of large numbers of nuclear weapons could have profound and severe effects on climate, especially in the Northern Hemisphere. This, in turn, could damage crops and potentially cause as much loss of life as the immediate effects of blast, thermal radiation, fire, and fallout. Several examples of devastating destruction are known throughout Earth's history. A 1980 paper in *Science* by Luis Alvarez, a Nobel Prize–winning physicist, and his colleagues described a global disaster 65 million years ago that they concluded resulted in huge decreases in temperature on Earth. They stimulated research on what R. P. Turco and his colleagues in 1983 termed "nuclear winter."

Alvarez and his colleagues sought to explain the sudden demise of the dinosaurs and many other species at the end of the Mesozoic era and the start of the Cenozoic geological era in Earth history. They proposed that Earth was hit by a large comet or asteroid that lofted debris high into the atmosphere and ignited multiple firestorms in forests. They sampled sediments at the Mesozoic-Cenozoic boundary at Gubbio, Italy, and reported unusually high concentrations of the element Iridium. Iridium, one of the platinum group of elements, is known to be more plentiful in meteorites than on earth. They concluded that the impact caused global decreases in temperature, loss of vegetation, and winter conditions that lasted long enough that many animals, such as dinosaurs, died of starvation.

The impact site was subsequently identified near the north coast of Mexico's Yucatán Peninsula. Dated sedimentary deposits along the northern Gulf of Mexico indicate that a huge sea wave (tsunami) hit those shores soon after the impact. Damage to living species was particularly high in North America to the northwest of the impact site, leading to the conclusion that the impacting body traveled in a northwesterly direction before hitting Yucatán.

Large volcanic eruptions, especially those that release large amounts of sulfur dioxide into the atmosphere, are known to cause drops in global temperature for one to a few years. The large eruption of the Indonesian volcano Krakatoa in 1883 caused global cooling of about 1.8° F (1° C) for two years. The even larger eruption of the earlier Indonesian volcano Tambora in 1815, one of the most powerful in recorded history, was followed by "the year without a summer." Temperature decreases after such large volcanic eruptions are useful tests of models of longer-term climate changes and the role of human-induced global warming.

I have selected one article that examines possible climatic changes associated with a major nuclear exchange between the Soviet Union and the United States and another that describes the effects of an exchange between India and Pakistan.

Scientists Turco, A. B. Toon, T. P. Ackerman, J. B. Pollack, and C. Sagan—referred to widely as TTAPS—made computer simulations for their 1983 article in *Science* of a number of scenarios involving the exchange of nuclear weapons between the Soviet Union and the United States. Their baseline exchange involved 10,400 explosions with yields of 100 to 10,000 kilotons (0.1 to 10 megatons [Mt]). Fortunately, this example is not valid today because bilateral treaties have reduced the numbers of warheads of the two nuclear superpowers (see figure 17.1). Also, the maximum yields of warheads decreased as more missiles with multiple warheads were deployed and bomber payloads transitioned from mostly high-yield bombs to many missiles with weapons of lower yield.

TTAPS scenarios 12 and 14 are more appropriate today. Case 12 involved an exchange of 2250 warheads with yields of 200 to 1000 kilotons; two-thirds were focused on hardened targets like missile silos and the remaining one-third on urban and industrial targets. Scenario 14 involved 1000 warheads with yields of 100 kilotons (ten times smaller than in case 12) used solely against urban and industrial targets. According to TTAPS, both of these scenarios would still have great long-term consequences. A major exchange aimed at military and industrial targets alone would not be just a "surgical strike" with damage limited to those facilities, because many are located within or near urban zones. In addition to radioactive fallout, widespread fires would occur after most nuclear bursts over forests and cities.

TTAPS also focused on the potential effects on climate of huge amounts of smoke and dust carried into the atmosphere by a major exchange, which would cause cooling of the lower atmosphere by blocking sunlight from reaching the surface of the Earth. Scenario 12, while less severe in its absolute impact than their baseline scenario, was projected to affect the atmosphere to an extent comparable to or exceeding that of a major volcanic eruption such as Tambora in 1815 and its following "year without a summer." They stated, "Unexpectedly, less than 1 percent of existing [1983] strategic arsenals, if targeted on cities, could produce optical (and climatic) disturbances much larger than those previously associated with a massive nuclear exchange ~10,000 MT." For case 12, 1 percent in 1983 is roughly equivalent to 10 percent of the reduced strategic arsenals of the United States and Russia in 2017.

The Earth's atmosphere consists of the troposphere, which extends from the surface to an altitude of about 6 to 8 miles (10 to 13 km). Above the troposphere is the stratosphere, which extends to an altitude of 30 miles (50 km).

TTAPS found that more than two to three months after a major nuclear exchange, while soot would be largely depleted in the atmosphere by rainfall and washout, dust would dominate optical effects. In case 14, none of the smoke from urban fires in about a hundred cities would reach the stratosphere, whereas for many of their other scenarios it would. TTAPS found that fine dust in the stratospheric would be responsible for prolonged cooling lasting a year or more.

TTAPS calculated average cooling of land areas in the Northern Hemisphere as large as about 58° F (32° C) for case 14, the city attack, and about 11° F (6° C) for case 12. This was so even though case 14 involved only 10 percent of the megatonnage of case 12. Significant temperature changes would last about a hundred days for case 14 and about twenty to eighty days for case 12. This would mean subfreezing temperatures in many places even in the summer for case 14. According to TTAPS, much of the population of the Earth might survive the immediate consequences of a nuclear war, but "the longer-term and global-scale aftereffects of nuclear war might prove to be as important as the immediate consequences of the war [i.e., the effects from blast, thermal radiation, and fallout]."

In the decades after these early 1980s, studies on nuclear winter, computer power, and global circulation models of the atmosphere and oceans improved immensely, with three-dimension models replacing one-dimensional ones. In 2007, some of the authors of the 1983 TTAPS study revisited the subject using a modern climate model with current nuclear arsenals. They reached similar conclusions about firestorms created by attacks on about a hundred cities (like case 14 of 1983) and calculated temperatures plunging below freezing in the summer in major agricultural regions, threatening food supplies for much of the planet. Climatic effects of smoke from burning cities and industrial areas, lofted into the upper stratosphere, would last for several years, longer than they originally thought in 1983.

Also in 2007, Toon, one of the TTAPS authors, and colleagues did a computer analysis of a major exchange of nuclear weapons between Pakistan and India. They assumed that one hundred Hiroshima-size weapons of about 15 kilotons each were used to attack the densest population centers in each country—generally huge megacities. They concluded that those explosions would generate substantial global-scale temperature anomalies, though not as large as in a major exchange between the United States and Russia. The effects would degrade agricultural productivity to an extent that has led historically to famines in Africa, India, and Japan.

Wikipedia describes work and speculation about the effects of the burning of oil wells in Kuwait that were ignited by Saddam Hussein of Iraq in 1991 during the first Gulf War. About 600 wells were ignited; some were not extinguished for more than six months. Prior to their being ignited, Turco, J. W. Birks, Carl Sagan, A. Robock, and P. Crutzen stated to reporters from two newspapers that they expected catastrophic nuclear winter effects if Iraq went through with its threats to ignite 300 to 500 oil wells and if they burned for a few months.

S. Fred Singer of the University of Virginia, a prominent denier of climate changes from human activities, and Sagan of Cornell University debated possible impacts of oil well fires in Kuwait on a TV news program. Sagan argued that some of the effects of the smoke lofting into the stratosphere could be similar to effects of nuclear winter and very similar to those from the eruption of Tambora in 1815. Singer, however, said his calculations showed that the smoke would rise to 3000 feet (900 m) and would be rained out in several days.

Sagan later conceded that his predictions did not turn out to be correct. He said, "it *was* pitch black at noon and temperatures dropped 7 to 11° F (4 to 6° C) over the Persian Gulf but not much smoke reached stratospheric [higher] altitudes and Asia was spared."

A 2007 study by Toon and others applied modern computer models to the Kuwait oil fires and found that individual smoke plumes were not able to loft smoke into the stratosphere. Nevertheless, smoke from fires over a larger area could extend into the stratosphere.

In the troposphere, the lower part of the atmosphere, temperature decreases with height; the troposphere turns over by convection and hence "washes itself out" with rain. The stratosphere, however, is more stratified (layered) because temperature there increases with height. Thus, small soot particles that make it into the stratosphere can remain there for a long time.

Criticisms of the "nuclear winter" concept and the effects on the atmosphere of the fires in Kuwait led many U.S. policy makers to ignore the possible consequences of a nuclear winter. One statement called it "nuclear fall." Note that the papers I have cited by Turco, Toon, and others were published in prominent refereed journals, whereas the remarks by Sagan and Singer were not. In 1987, Michael Kelly of Cambridge University and the British Climatic Research Unit stated, "although there are a handful of vociferous critics, the atmospheric community is united in its conclusion that the threat of nuclear winter is genuine."

After the breakup of the Soviet Union in 1989, relatively little work was published on nuclear winter. I strongly believe a vigorous debate on the subject is needed today. Even without as severe climatic effects as proposed by TTAPS in 1983, it is hard to dismiss the huge climatic effects from a major nuclear exchange.

ACCOMPLISHMENTS OF SEISMOLOGISTS IN MONITORING A FULL TEST BAN TREATY

Since the first calls for a CTBT in the 1950s, seismological instrumentation and techniques to monitor nuclear testing have improved immensely. None of the states possessing nuclear weapons that signed the CTBT in 1996 has detonated a nuclear explosion of military significance for twenty years. The CTBT has acted as a barrier to the development and testing of new generations of nuclear weapons. Serious evasion schemes, while still topics of political and occasional scientific debate, are considered exceedingly difficult to conduct without being detected by seismic methods, radionuclide sampling, and satellite imagery. Methods to muffle nuclear explosions in underground cavities have been determined to be unfeasible down to very small yields. Over the past fifty-five years since I became a graduate student at Columbia, the field of seismology has come full circle, finally fulfilling its long-thwarted promise to verify a nuclear test ban.

PSYCHOLOGICAL ASPECTS OF THE NUCLEAR ARMS RACE

I have read a number of publications by two psychiatrists, Robert Jay Lifton and Jerome Frank, who have written on psychological aspects of the nuclear arms race. I was fortunate to meet briefly with Frank at Johns Hopkins University when I gave an invited lecture on verifying a full test ban treaty. Frank, who died in 2005, was one of the founders of Physicians for Social Responsibility. I recommend his 1982 book *Sanity and Survival in the Nuclear Age: Psychological Aspects of War and Peace.*

Following his work on Hiroshima survivors, Lifton became a vocal opponent of nuclear weapons. He argued that nuclear strategy and war-fighting doctrines made even mass genocide banal and conceivable.

Among his books, I recommend *The Broken Connection: On Death and the Continuity of Life* (Simon & Schuster, 1979) and *Indefensible Weapons: The Political and Psychological Case against Nuclearism* (Basic Books, 1982). Very little attention is devoted to this important topic in the United States.

WHY I WORK ON SUCH A FRIGHTENING TOPIC

I am sometimes asked why I work on such a frightening and depressing topic. I explain to myself that this is *the* major issue of my lifetime. With my scientific knowledge, I hope to contribute in some small way to preventing the use of nuclear weapons. I regard this as my duty as an informed citizen, especially in a country that possesses vast numbers of nuclear weapons. I hope this book will convince others to learn more about these issues and to become more involved. I support the advice of Edmund Burke, the British-Irish orator, political theorist, and philosopher, who said, "Nobody made a greater mistake than the one who did nothing because they could only do a little."

A major nuclear exchange would be a cataclysmic disaster with a level of destruction unprecedented in the entire history of our species. Some people have argued that because nuclear weapons have not been used since 1945, the probability of their use is very small. The world has been fortunate that nuclear weapons have not been used since then, but this could end at a moment's notice. False alarms, accidents, and the near miss of the Cuban missile crisis are not very reassuring about nuclear weapons' not being used in the future. The probability per year of a nuclear exchange may be low, but if it happens, the consequences will be catastrophic. Getting the public and governments to deal with rare but catastrophic events is difficult but very necessary.

The Trump administration has made threatening remarks about nuclear weapons. As of mid-2017 it is not clear if it might either use nuclear weapons against an advisory such as North Korea or resume nuclear testing. If it resumed testing, the yields of explosions likely would be large, abrogating several arms control agreements, and other countries almost certainly would resume testing.

FIGURE 18.1

The Doomsday Clock of the *Bulletin of the Atomic Scientists*. October 2016.

Since 1947, the *Bulletin of the Atomic Scientists* has published a Doomsday Clock symbolizing the dangers to humanity of a nuclear exchange (figure 18.1). It has been set between two and seventeen minutes to midnight at various times since 1947. In early 2017 it was reset from three to two and a half minutes to midnight.

GLOSSARY AND ABBREVIATIONS

ABM (ANTI-BALLISTIC MISSILE)—A missile to knock down other missiles

ABM TREATY—Treaty between United States and USSR on limitation of ABMs

AEDS (ATOMIC ENERGY DETECTION SYSTEM)—Operated by AFTAC

AFTAC (AIR FORCE TECHNICAL APPLICATIONS CENTER)—Does classified monitoring

BMD (BALLISTIC MISSILE DEFENSE)—Proposals to destroy incoming missiles with missiles

BOOSTED WEAPON—Addition of hydrogen isotopes to a weapon to make it lighter

CD—UN's Conference on Disarmament

CTBT (COMPREHENSIVE TEST BAN TREATY)—Ban on testing of nuclear explosions of all yields

CTBTO (COMPREHENSIVE TEST BAN TREATY ORGANIZATION)—UN Agency for CTBT

CUBAN MISSILE CRISIS—Attempt by USSR to introduce nuclear weapons into Cuba in October 1962 and U.S. response

DECOUPLED NUCLEAR TEST—Test fired in a large underground cavity so as to reduce the size of seismic waves produced

EPICENTER—Location of seismic event in latitude and longitude

EVERNDEN, JACK—Seismologist involved with test bans

FAT MAN—Weapon tested in New Mexico and dropped on Nagasaki in 1945

FISSILE MATERIAL—Certain isotopes of uranium and plutonium capable of undergoing nuclear fission

FISSION—Splitting heavier into lighter elements

FUSION—Fusing hydrogen isotopes into helium

HERRIN, GENE—Seismologist involved with nuclear tests

HEU (HIGHLY ENRICHED URANIUM)—High percentage of uranium 235

HYDROACOUSTIC MONITORING—Detecting sound waves propagated in water

HYPOCENTER—Location of seismic event in three dimensions

IAEA (INTERNATIONAL ATOMIC ENERGY AGENCY)—UN agency based in Vienna

ICBM—Intercontinental ballistic missile

IMS (INTERNATIONAL MONITORING SYSTEM)—Monitors CTBT internationally

INF (INTERMEDIATE-RANGE NUCLEAR FORCES) TREATY—Treaty between United States and USSR for systems capable of delivering nuclear weapons between short and intercontinental distances

INFRASOUND MONITORING—Detecting very low frequency sound waves propagated in the atmosphere

ISOTOPES—Varieties (flavors) of an element with different numbers of neutrons but the same number of protons in the atomic nucleus

KAZAKHSTAN—Country in Central Asia, formerly part of USSR

KIRGHIZSTAN—Country in Central Asia, formerly part of USSR

KT (KILOTON)—Energy equivalent to 1000 tons of TNT

LEP (LIFE EXTENSION PROGRAM)—Longer lifetime for a U.S. nuclear weapon

LITTLE BOY—Weapon dropped on Hiroshima in 1945

LOP NOR—Nuclear test site of China

LRD (LONG-RANGE DETECTION)—Operated by U.S. Air Force to monitor nuclear testing by other countries

LTBT (LIMITED TEST BAN TREATY)—1963 treaty that banned nuclear tests in the atmosphere, in space, and underwater

MAGNITUDE—A measure of the size of a seismic event

MICROEARTHQUAKES—Earthquakes smaller than magnitude 3

MIRV—Multiple, independently targetable reentry vehicle containing two or more warheads

MT (MEGATON)—Energy equal to 1 million tons of TNT

MW (MOMENT MAGNITUDE)—Very long period of seismic magnitude

NGO—A nongovernmental organization

NIF (NATIONAL IGNITION FACILITY)—Lasers at Livermore used to create fusion

NOVAYA ZEMLYA—Arctic test sites of Soviet Union

NPT (NUCLEAR NONPROLIFERATION TREATY)—Multilateral treaty on possession and development of nuclear weapons

NTM (NATIONAL TECHNICAL MEANS)—A nation's monitoring facilities

NTS—Nevada Test Site of the United States

NUCLEAR ARMS CONTROL—Limits or restraints on nuclear weapons and/or their delivery vehicles that are mutually agreed upon between states

NWFZ (NUCLEAR WEAPONS FREE ZONE)—A region agreed by states within it to be nuclear free by treaty

OPPENHEIMER, J. ROBERT—Director of Los Alamos during World War II

PNET (PEACEFUL NUCLEAR EXPLOSIONS TREATY)—Prohibition on testing of peaceful nuclear explosions underground larger than 150 kilotons by USSR and United States

PRESS, FRANK—Geophysicist; President Carter's science adviser; president of the National Academy of Sciences

RADIONUCLIDE MONITORING—Technologies to detect radioactive particulates and noble gases

RICHARDS, PAUL—Seismologist involved with test bans

RICHTER, CHARLES—Seismologist who devised first magnitude scale

ROMNEY, CARL—Seismologist involved with test bans

SALT I AND II (STRATEGIC ARMS LIMITATION TREATIES)—Limitations on intercontinental delivery systems agreed to by United States and USSR

SEABORG, GLEN—Scientist who discovered plutonium; high-level adviser during Kennedy and Johnson administrations

SEISMICITY—Descriptions of locations and sizes of earthquakes and their relationship(s) to geological features

SEISMOLOGY—Study of earthquakes and earth structure

SORT (STRATEGIC OFFENSIVE REDUCTION TREATY)—Follow-on to SALT treaties

START (STRATEGIC ARMS REDUCTION TREATIES)—Follow-on to SALT

STOCKPILE STEWARDSHIP—A program to ensure that nuclear weapons will work in the future without nuclear testing

TECTONICS—Discipline of geology involving processes that control structure, deformation, and properties of Earth, moons, and planets

TELLER, EDWARD—Physicist involved with development of U.S. nuclear weapons

THERMONUCLEAR WEAPON—Sometimes called a hydrogen or fusion bomb; fusion only takes place at temperatures of millions of degrees, requiring triggering by a primary fission explosion

TRIAD—Delivery system consisting of land- and sea-based missiles and aircraft

TTBT (THRESHOLD TEST BAN TREATY)—Prohibition on underground testing of nuclear weapons larger than 150 kilotons agreed to by USSR and United States

USSR—Union of Soviet Socialist Republics

WARHEAD—The nuclear weapon carried by a delivery system

YIELD—Energy released by a nuclear explosion

REFERENCES

Air Force Technical Applications Center (AFTAC). *A Fifty Year Commemorative History of Long Range Detection.* Patrick Air Force Base, FL: Air Force Technical Applications Center, 1997.

Albright, D., P. Brannan, Z. Laporte, K. Tajer, and C. Walrond. *Rendering Useless South Africa's Nuclear Test Shafts in the Kalahari Desert.* Institute for Science and International Security, November 30, 2011.

Alvarez, L. W., W. Alvarez, F. Asaro, and H. V. Michel. "Extraterrestrial Cause for the Cretaceous–Tertiary Extinction." *Science* 208 (1980): 1095–1108.

Ambraseys, N. N., and C. P. Melville. *A History of Persian Earthquakes.* London: Cambridge University Press, 1982.

Bache, T. C. *Estimating the Yield of Underground Nuclear Explosions.* DARPA–Geophysical Sciences Division 82–03, 1982.

Bache, T. C., and R. W. Alewine. "Monitoring a Comprehensive Test Ban Treaty." Presentation to the American Geophysical Union, Special Session on Verification of Nuclear Test Bans, Baltimore MD, June 2, 1983.

Bache, T. C., W. J. Best, R. R. Blandford, G. V. Bulin, D. G. Harkrider, E. J. Herrin, A. Ryall, and M. J. Shore. *A Technical Assessment of Seismic Yield Estimation.* DARPA NMR 81–02, 1981.

Bache, T. C., S. R. Bratt, and L. B. Bache. "P Wave Attenuation, m_b Bias and the Threshold Test Ban Treaty." *DARPA Annual Review*, 1986, 194–201.

Bailey, K. C. *Hydronuclear Experiments: Why They Are Not a Proliferation Danger.* UCRL-ID-118538. Livermore, CA: Lawrence Livermore National Lab, 1994.

Berberian, M. "100 Years; 126,000 Deaths." Encyclopaedia Iranica. Abadan Publishing Co., 1997. http://iranian.com/Iranica/June97/Earthquake/Text2.html.

Blair, B. *The Logic of Accidental Nuclear War.* Washington, DC: Brookings Institution, 1993.

Blandford, R. R., R. H. Shumway, R. Wagner, and K. L. McLaughlin. *Magnitude Yield for Nuclear Explosions at Several Test Sites with Allowance for Effects of Truncated Data, Amplitude Correlation Between Events Within Test Sites, Absorption and pP.* Technical Report, TGAL-TR-83-6, Teledyne Geotech, 1984.

Bolt, B. A. *Nuclear Explosions and Earthquakes: The Parted Veil*. San Francisco: Freeman, 1976.

Carothers, J. E. *Caging the Dragon: The Containment of Underground Nuclear Explosions*. U.S. Department of Energy Nuclear Agency Report DOE/NV-388 DNA TR 95–74. Washington, DC: 1995.

Cochran, T. B., W. M. Arkin, and M. M. Hoenig. *U.S. Nuclear Forces and Capabilities: Vol. I. Nuclear Weapons Databook*. Cambridge, MA: Ballinger, 1984.

Cochran, T. B., W. M. Arkin, R. S. Norris, and J. I. Sands. *Soviet Nuclear Weapons: Vol. IV. Nuclear Weapons Databook*. New York: Harper & Row, 1989.

Collina, T. Z., and D. G. Kimball. *Now More Than Ever: The Case for the Comprehensive Nuclear Test Ban Treaty*. Washington DC: Arms Control Association, 2010.

Dahlman, O., and H. Israelson. *Monitoring Underground Nuclear Explosions*. Amsterdam: Elsevier, 1977.

Dahlman, O., S. Mykkeltveit, and H. Haak. *Nuclear Test Ban: Converting Political Visions to Reality*. Springer, 2009.

Davis, D. M., and L. R. Sykes. "Geological Constraints on Clandestine Nuclear Testing in South Asia." *Proceeding of the U.S. National Academy of Sciences* 96 (1999): 11090–11095.

Engdahl, E. R., J. A. Jackson, S. C. Myers, E. A. Bergman, and K. Priestley. "Relocation and Assessment of Seismicity in the Iran Region." *Geophysical Journal International* 167 (2006): 761–778.

Evernden, J. F.. "Magnitude Determination at Regional and Near-Regional Distances in the United States." *Bulletin of the Seismological Society America* 57 (1967): 591–639.

Evernden, J. F., and L. R. Sykes. "Nuclear Test Yields." *Science* 223 (1984): 642–644.

Feiveson, H. A., A. Glaser, Z. Mian, and F. N. Von Hippel. *Unmaking the Bomb: A Fissile Material Approach to Nuclear Disarmament and Nonproliferation*. Cambridge, MA: MIT Press, 2014.

Fisk, M. D., and W. S. Phillips. "Constraining Regional Phase Amplitude Models for Eurasia: Part 2. Frequency-Dependent Attenuation and Site Effects." *Bulletin of the Seismological Society America* 103 (2013): 3265–3288.

Hersh, S. M. *The Sampson Option: Israel's Nuclear Arsenal and American Foreign Policy*. New York: Random House, 1991.

Heuze, F. E., D. R. Wilburn, J. A. Russell, and D. I. Bleiwas. *Estimated Use of Explosives in the Mining Industries of Algeria, Iran, Iraq, and Libya*. Lawrence Livermore National Laboratory, September 1995.

Khalturin, V. I., T. G. Rautian, P. G. Richards, and W. S. Leith. "A Review of Nuclear Testing by the Soviet Union at Novaya Zemlya, 1955–1990." *Science and Global Security* 13 (2005): 1–42.

Kidder, R. E. "Militarily Significant Nuclear Explosive Yields." *F.A.S. Public Interest Report: Journal of the Federation of American Scientists (FAS)* 38, no. 7 (1985): 1–3.

Kim, W.-Y., P. G. Richards, and L. R. Sykes. "Discrimination of Earthquakes and Explosions Near Nuclear Test Sites Using Regional High-Frequency Data." Poster

SEISMO-27J. International Scientific Studies Conference, Comprehensive Test Ban Treaty Organization, Vienna, June 2009.

Kulp, L. R., A. R. Schubert, and E. J. Hodges. "Strontium-90 in Man III: The Annual Increase of This Isotope and Its Pattern of World-Wide Distribution in Man Defined." *Science* 129 (1959): 1249–1255.

Kværna, T., F. Ringdal, J. Schweitzer, and L. Taylor. "Optimized Seismic Threshold Monitoring—Part 1: Regional Processing." *Pure and Applied Geophysics* 159 (2002): 969–987.

Marshall, P. D., and D. L. Springer. "Is the Velocity of P_n an Indicator of Q_α?" *Nature* 264 (1976): 531–533.

Marshall, P. D., D. L. Springer, and H. C. Rodean. "Magnitude Corrections for Attenuation in the Upper Mantle." *Geophysical Journal of the Royal Astronomical Society* 57 (1979): 609–638.

Mikhailov, V. N., ed. *USSR Nuclear Weapons Tests and Peaceful Nuclear Explosions: 1949 Through 1990*. Sarov, Russia: Ministry of the Russian Federation for Atomic Energy and Ministry of Defense of the Russian Federation, 1996.

National Academies of the United States. *The Comprehensive Nuclear Test Ban Treaty—Technical Issues for the United States*. Washington, DC: National Academies Press, 2012. Previous study published in 2002.

Office of Technology Assessment (OTA), Congress of the United States. *The Containment of Underground Nuclear Explosions*. Washington, DC: U.S. Government Printing Office, 1989.

——. *Seismic Verification of Nuclear Testing Treaties*. Washington, DC: U.S. Government Printing Office, 1988.

Pasyanos, M. E., E. M. Matzel, W. R. Walter, and A. J. Rodgers. "Broad-Band Lg Attenuation Modelling in the Middle East." *Geophysical Journal International* 177 (2009): 1166–1176.

Pike, J., and J. Rich. "The Threshold Test Ban Treaty (TTBT): Have the Soviets Exceeded the 150 Kiloton Limit?" *F.A.S. Public Interest Report: Journal of the Federation of American Scientists (FAS)* 37, no. 3 (1984): 16–20.

Reed, T. C., and D. B. Stillman. *The Nuclear Express*. Minneapolis MN: Zenith, 2009.

Rhodes, R. *Dark Sun: The Making of the Hydrogen Bomb*. New York: Simon and Schuster, 1995.

——. *The Making of the Atomic Bomb*. New York: Simon and Schuster, 1986.

Ringdal, F., P. D. Marshall, and R. W. Alewine. "Seismic Yield Determination of Soviet Underground Nuclear Explosions at the Shagan River Test Site." *Geophysical Journal International* 109 (1992): 65–77.

Robinson, C. P., J. Foster, and T. Scheber. "The Comprehensive Test Ban Treaty: Questions and Challenges." Lecture delivered April 10, 2012. *The Heritage Foundation*, no. 1218 (November 7, 2012).

Romney, C. *Detecting the Bomb: The Role of Seismology in the Cold War*. Washington, DC: New Academic, 2009.

Seaborg, G. T. *Kennedy, Khrushchev and the Test Ban*. Berkeley: University of California Press, 1981.

Shultz, G. P., S. P. Andreasen, S. P. Drell, and J. E. Goodby, eds. *Reykjavik Revisited: Steps Toward a World Free of Nuclear Weapons*. Stanford, CA: Hoover Institution Press, Stanford University, 2008.

Shultz, G. P., W. J. Perry, H. A. Kissinger, and S. Nunn. "Toward a Nuclear-Free World." *Wall Street Journal*, January 4, 2008, A15.

Springer, D. L., G. A. Pawloski, J. L. Ricca, R. F. Rohrer, and D.K. Smith. "Seismic Source Summary for All U.S. Below-Surface Nuclear Explosions." *Bulletin of the Seismological Society of America* 92 (2002): 1806–1840.

Starr, S., L. Eden, and T. A. Postol. "What Would Happen If an 800-Kiloton Nuclear Warhead Detonated Above Midtown Manhattan?" *Bulletin of the Atomic Scientists* 71 (2015).

Sultanov, D. D., J. R. Murphy, and Kh. D. Rubinstein. "A Seismic Source Summary for Soviet Peaceful Nuclear Explosions." *Bulletin of the Seismological Society of America* 89 (1999): 640–647.

Sykes, L. R. "Dealing with Decoupled Nuclear Explosions Under a Comprehensive Test Ban Treaty." In *Monitoring a Comprehensive Nuclear Test Ban Treaty*, ed. E. S. Husebye and A. M. Dainty, 247–293. Dordrecht, Netherlands: Kluwer Academic, 1996.

——. "False and Misleading Claims About Verification During the Senate Debate on the Comprehensive Nuclear Test Ban Treaty." *F.A.S. Public Interest Report: Journal of the Federation of American Scientists (FAS)* 53, no. 3 (2000): 1–12.

——. "Four Decades of Progress in Seismic Identification Help Verify the CTBT." *Eos, Transactions American Geophysical Union* 83, no. 44 (2002): 497–500.

——. "Small Earthquake Near Russian Test Site Leads to U.S. Charges of Cheating on Comprehensive Nuclear Test Ban Treaty." *F.A.S. Public Interest Report: Journal of the Federation of American Scientists (FAS)* 50, no. 6 (1997): 1–12.

Sykes, L. R., and D. M. Davis. "The Yields of Soviet Strategic Weapons." *Scientific American* 256, no. 1 (1987): 29–37.

Sykes, L. R., and J. F. Evernden. "The Verification of a Comprehensive Nuclear Test Ban." *Scientific American* 247, no. 4 (1982): 47–55.

Sykes, L. R., J. F. Evernden, and I. Cifuentes. "Seismic Methods for Verifying Nuclear Test Bans." In *Physics, Technology, and the Nuclear Arms Race*, ed. D. W. Hafemeister and D. Schroeer, 85–133. New York: American Institute of Physics, 1983. Figures and one table reproduced with the permission of AIP Publishing.

Sykes, L. R., and M. Nettles. "Dealing with Hard-to-Identify Seismic Events Globally and Those Near Nuclear Test Sites." Poster SEISMO 26J. International Scientific Studies Conference, Comprehensive Nuclear Test Ban Treaty Organization, Vienna, June 2009.

Toon, O. B., A. Robock, R. P. Turco, C. Bardeen, L. Oman, and G. L. Stenchikov. "Consequences of Regional-Scale Nuclear Conflicts." *Science* 315 (2007): 1224–1225.

Turco, R. P., O. B. Toon, T. P. Ackerman, J. B. Pollack, and C. Sagan. "Nuclear Winter: Global Consequences of Multiple Nuclear Explosions." *Science* 222 (1983): 1283–1292.
Von Hippel, F.. "Gorbachev's Unofficial Arms-Control Advisers." *Physics Today*, September 2013, 41–47.
Weinberg, S. "Can Missile Defense Work?" *New York Review of Books*, February 14, 2002, 41–48.

Note: Ron Doel, now of Oregon State University, recorded seven sessions of oral history with me in 1996 and 1997. They include my childhood, education, the development of plate tectonics, and my work on nuclear test verification. Each is available through either the Oral History Research Office of Columbia University or the American Institute of Physics. I was not given the opportunity to correct spelling and other minor mistakes in the AIP transcript before it was placed on their website.

INDEX

ABOUT THE AUTHOR

Dr. Lynn Sykes has been involved in the identification of underground nuclear testing and the long battle to obtain a total ban on nuclear testing for the past fifty-five years. In 1986 the Federation of American Scientists presented Sykes and two colleagues with its Public Service Award for "Leadership in Applying Seismology to the Banning of Nuclear Tests, Creative in Utilizing Their Science, Effective in Educating Their Nation, Fearless and Tenacious in Struggles within the Bureaucracy." He also received the John Wesley Powell Award from the U.S. Geological Survey for work on U.S. earthquakes in 1991.

Sykes became a member of the Staff of the Lamont-Doherty Earth Observatory of Columbia University in 1965, remaining there until 2005 when he retired as a professor emeritus. After graduating with bachelors and masters degrees in geology from MIT in 1960, he earned his Ph.D. in seismology at Columbia University in 1965. He became a faculty member in 1968 and then the Higgins Professor of Earth and Environmental Sciences.

In 1974, as a leading seismologist, Dr. Sykes was invited to become a member of the U.S. delegation that traveled to the Soviet Union to negotiate the Threshold Test Ban Treaty. He testified before the U.S. Congress numerous times as an expert on nuclear-test verification, a subject with large scientific and public policy components.

Sykes, along with Walter Pitman of Lamont and Jason Morgan of Princeton, showed unequivocally that the Earth's outermost layers consist of nearly rigid plates that move over the surface. Referred to as plate tectonics, it revolutionized the study of the Earth's crust, providing an understanding of the formation of mountain ranges, the drifting of the continents, volcanoes, earthquakes, ocean basins, mid-oceanic ridges, deep sea trenches, the evolution of climate, and the distribution of natural resources. Sykes' research illustrated the importance of great faults that intersect mid-ocean ridges in accommodating plate motion and on underthrusting of plates at subduction zones. Dr. Marcia McNutt, a geophysicist and President of the U.S. National Academy of Sciences, called the discovery of plate tectonics "one of the top ten scientific accomplishments of the second half of the 20th century." The three scientists were awarded the prestigious Vetlesen Prize in 2000.

Sykes is a member of the National Academy of Sciences and the American Academy of Arts and Sciences and a Fellow of the Geological Society of America and of the American Geophysical Union, which honored him with its Macelwane and Bucher awards. He also received the Seismological Society of America's most prestigious H. F. Reid medal. While officially retired, he continues his research on earthquakes and nuclear explosions and has been hard at work writing this book and one on plate tectonics and great earthquakes.